elementals

v. an *elemental* life

volume v
an *elemental* life

John Hausdoerffer, editor
Nickole Brown & Craig Santos Perez, poetry editors

Gavin Van Horn & Bruce Jennings, series editors

Humans and Nature Press, Libertyville 60030
© 2024 by Center for Humans and Nature

For more information, contact Humans & Nature Press,
17660 West Casey Road, Libertyville, Illinois 60048.
Printed in the United States of America.

Cover and slipcase design: Mere Montgomery of LimeRed, https://limered.io

ISBN-13: 979-8-9862896-3-2 (paper)
ISBN-13: 979-8-9862896-4-9 (paper)
ISBN-13: 979-8-9862896-5-6 (paper)
ISBN-13: 979-8-9862896-6-3 (paper)
ISBN-13: 979-8-9862896-7-0 (paper)
ISBN-13: 979-8-9862896-2-5 (set/paper)

Names: Hausdoerffer, John, editor | Brown, Nickole, poetry editor | Perez, Craig Santos, poetry editor | Van Horn, Gavin, series editor | Jennings, Bruce, series editor

Title: Elementals: An Elemental Life, vol. 5 / edited by John Hausdoerffer

Description: First edition. | Libertyville, IL: Humans and Nature Press, 2024 | Identifiers: LCCN 2024902609 | ISBN 9798986289670 (paper)

Copyright and permission acknowledgments appear on pages 168–169.

Humans and Nature Press
17660 West Casey Road, Libertyville, Illinois 60048

www.humansandnature.org

Printed by Graphic Arts Studio, Inc. on Rolland Opaque paper. This paper contains 30% post-consumer fiber, is manufactured using renewable energy biogas, and is elemental chlorine free. It is Forest Stewardship Council® and Rainforest Alliance certified.

contents

Gathering: Introducing the Elementals Series

Gavin Van Horn and Bruce Jennings

Thunderous, cymbal-clashing waves. Dervish winds whipping across mountain saddles. Conflagrations of flame licking at a smoke-filled sky. The majesties of desert sands and wheat fields extending beyond the horizon. What riotous confluence of sound, sight, smell, taste, and touch breaches your imagination when you call to mind the elementals? Yet the elementals may enter your thoughts as subtler, quieter presences. The gentle burbling of clear creek water. The rich loamy soil underfoot on a trail not often followed. A pine-scented breeze wafting through a forest. The inviting warmth of a fire in the hearth.

This last image of the hearth fire is apropos for the five volumes that constitute *Elementals*. The fire, with its gift of collective warmth, is a place to gather and cook together, and not least of all a place that invites storytelling. And in stories the elementals can be imagined as a better way of living still to be attained.

The essays and poems in these volumes offer a wide variety of elemental experiences and encounters, taking kaleidoscopic turns into the many facets of earth, air, water, and fire. But this series ventures beyond good storytelling. Each of the contributions in the pages you now hold in your hands also seeks to respond to a question: What can the vital forces of earth, air, water, and fire teach us about being human in a more-than-human world? Perhaps this sort of question is also part of experiencing a good fire, the kind in which we can stare into the sparks and contemplate our lives, releasing our imaginations to possibilities, yet to be fulfilled but still within reach. The elementals live. Thinking and acting through them—in accommodation with them—is not outmoded in our time. On the contrary, the rebirth of elemental living is one of our most vital needs.

For millennia, conceptual schemes have been devised to identify and understand aspects of reality that are most essential. Of enduring fascination are the four material elements: earth, air, water, and fire. For much longer than humans have existed—indeed, for billions of years—the planet has been shaped by these powerful forces of change and regeneration. Intimately part of the geophysical fluid dynamics of the Earth, all living systems and living beings owe their existence and well-being to these elemental movements of matter and flows of energy. In an era of anthropogenic influence and climate destabilization, however, we are currently bearing witness to the dramatic and destructive potential of these forces as it manifests in soil loss, rising sea levels, devastating floods, and unprecedented fires. The planet absorbs disruptions brought about by the activity of living systems, but only within certain limits and tolerances. Human beings collectively have reached and are beginning to exceed those limits. We might consider these events, increasing in frequency and intensity, as a form of pushback from the elementals, an indication that the scale and scope of human extractive behaviors far exceeds the thresholds within which we can expect to flourish.

The devastating unleashing of elemental forces serves as an invocation to attend more deeply to our shared kinship with other creatures and to what is life-giving and life-nurturing over long-term time horizons. In short, caring about the elementals may also mean caring for them, taking a more care-full approach to them in our everyday lives. And it may mean attending more closely to the indirect effects of technological power employed at the behest of rapacious desires. Unlike more abstract notions of nature or numerical data about species loss, air measurements in parts per million, and other indicators of fraying planetary relations, the elementals can ground our moral relations in something tangible and close at hand—near as our next breath, our next meal, our next drink, our next dark night dawning to day.

For each element, the contributors to this series—drawing from their diverse geographical, cultural, and stylistic perspectives—explore and illuminate practices and cosmovisions that foster reciprocity between people and place, human and nonhuman kin, and the living energies that make all life possible. The essays and poems in this series frequently approach the elements from unexpected angles—for example, asking us to consider the elemental qualities of bog songs, the personhood of rivers, yogic breath, plastic fibers, coal seams, darkness and bird migration, bioluminescence, green burial and mud, the commodification of oxygen, death and thermodynamics, and the healing sociality of a garden, to name only a few of the creatively surprising ways elementals can manifest.

Such diverse topics are united by compelling stories and ethical reflections about how people are working with, adapting to, and cocreating relational depth and ecological diversity by respectfully attending to the forces of earth, air, water, and fire. As was the case in the first anthology published by the Center for Humans and Nature, *Kinship: Belonging in a World of Relations,* the fifth and final volume of *Elementals* looks to how we can live in right relation, how we can *practice* an elemental life. There you'll find the elements converging in provocative ways, and sometimes challenging traditional ideas about what the elements are or can be in our lives. In each of the volumes of *Elementals,* however, our contributors are not simply describing the elementals; they are also always engaging the question, How are we to live?

In a sense, as a collective chorus of voices, *Elementals* is a gathering; we've been called around the fire to tell stories about what it means to be human in a more-than-human world. As we stare into this firelight, recalling and hearing the echoing voices of our living planet, we stretch our natural and moral imaginations. Having done so, we have an opportunity to think and experiment afresh with how to live with the elementals as good relatives. The elementals set the thresholds; they give feedback. Wisdom—if defined as thoughtful, careful practice—entails conforming to what the elements are "saying" and then learning (over a lifetime) how to better listen and respond. Pull up a chair, or sit on the ground near the crackling glow; we'll gaze into the fire together and listen—to the stories that shed light and comfort, to the stories that discombobulate and help us see old things in a new way, to the stories that bring us back to what matters for carrying on together.

Introduction: At the Headwaters of the Elemental Life

John Hausdoerffer

Nambé Falls, Nambé Pueblo, New Mexico invites us to accept, even to celebrate, the power of paradox—the same water that has gently nurtured beans for hundreds of years has also carved the deep rock canyon below. In April 2022, I traveled to Nambé Pueblo with a group of public lands management graduate students from Western Colorado University, to learn from pueblo communities about their land use strategies. Nambé Lieutenant Governor Nat Herrera taught us about their approaches to fire, gravity-fed water conservation, electric car innovation, and wildlife management. Herrera and his staff showed us how Nambé human and ecological health have always relied on waters that begin with the snows of twelve-thousand-foot Rock Man Mountain (which present-day maps call Santa Fe Baldy) towering above the pueblo, which melt into acequia irrigation systems that saturate Puebloan food systems, and eventually flow into the Rio Grande itself. Nambé Falls stands at the nucleus of this intricate ecological, cultural, political, economic, and cosmological system.

Herrera led my class to the falls. The sandy trail snaked up the mountain through pungent piñon, leading to a fifteen-foot-wide, hundred-foot-high canyon of cold, forty-five-degree water. I ventured off on my own, suspecting the students wanted space to ask Herrera questions. As I walked upstream, the slot canyon's sounds were deafening. Snowmelt from high above surged through the sandstone bottleneck of the falls where the canyon ended. I reached the bottom of the falls and found a rock lit by a shot of sunshine that slipped through the canyon. What a welcome place to sit: my upper body cooled from the mist of the falls as my lower body thawed in the sun after the trudge through snowmelt water. I relaxed, looking up at the falls. The elements—the rocky earth, the spray of the water, the fire of the sun, and the clarity of the high-altitude air—welcomed this solitary moment of reflection.

I lamented how rare these moments of kinship with the elements can be. I thought about how our relationship with the elements is in crisis on a planetary scale. In fact, April 2022 in New Mexico's Sangre de Cristo Mountains brought humbling reminders of our need to live in kinship with the elements. Snowpack, and thus water quantities, from the previous winter were at about 50 percent of normal levels; the air held April heat as if it were midsummer, a result of average New Mexican temperatures rising three degrees Fahrenheit in the past fifty years; the earth was dried up in anticipation of a fire season that would ultimately burn nine hundred thousand acres by the end of 2022 in New Mexico alone— more than three times the state's average over the previous century.

By the end of my class's week in the region, we looked at the eleven-thousand-foot ridge above Santa Fe to see the year's first plume of wildfire smoke in a dry, north-facing part of the Pecos Wilderness that regularly had snowpack long into May. I even remember, on a Memorial Day backpacking trip in the late 1990s, wishing I had snowshoes in those same Rock Man Mountain subalpine forests. Back then, I was a graduate student in Santa Fe who could only imagine climate impacts from seemingly distant data. Now that same section of forest is on fire, six weeks earlier than when I fell knee-deep into snow with each step up the mountain. Now, the data were felt and seen in everything around us as we toured the region, listening to forest ecologists, acequia farmers, environmental justice activists, and Nambé land managers.

On the day before we visited Nambé Falls, a local scientist showed us a project for diverting Rio Grande water to Santa Fe.

The waist-high, circular, concrete water tap was at least ten feet in diameter—like some dystopian megastraw along the streambank, sucking the region's lifeblood to sustain (maybe) an extra hundred years of excess consumption miles away. The scientist paraphrased a Puebloan elder, who told him that he could tell from that plan that the Western way of life will not last. Referring to a more patient pueblo sense of cyclical time and a continuing history of adapting to cultural and ecological shock, the elder said, "We're just waiting you out." The Anthropocene presents haunting lessons in humility, calling us beyond resource conservation to return to elemental kinship.

I did not know what to do with that paradox between the tenderness with which the elements sustained my being at the base of Nambé Falls and the severity with which the elements rejected my society's refusal to live in kinship with those same elements. Perhaps the former is key to living through the latter. Perhaps *living elementally*—living connected to and guided by the elements—can lead to making the elements more livable again.

This volume, *An Elemental Life*, is the final volume of the *Elementals* series—*Earth, Air, Water, Fire*. The purpose of this volume is to explore how a rethinking of the elements might inspire us to rethink how we live. If the elements are kin to one another, then what does it mean to live in kinship with the elements? If connectivity is supplanting atomization in our view of the elements, and if kinship is replacing individuality as our frame for living a good life, then how are the elements—the interconnected ingredients of our interstellar sphere of kin—telling us to live? This volume has a special place in the series because it brings the elements into conversation with one another by asking the practice-oriented question, What is an elemental life?

This volume is fortunate to have such an array of thoughtful writers join the conversation. Special thanks go to our poetry editors—Nickole Brown and Craig Santos Perez—for bringing such stirring verse to these pages. The poets Heather Swan, Allison Adelle Hedge Coke, Elizabeth Coleman, Matthew Olzmann, Sean Hill, Brenda Hillman, and CMarie Fuhrman collectively enliven Fuhrman's hope for an elemental life grounded in a "mutual recognition" between the lives of all beings and rooted in an understanding that our "Earthly purpose" is the "mere act of giving / love to that which we took no part in creating."

The essays and interviews were also an honor to gather. Although not official sections of the volume, I have divided the authors into three layers of the elemental life: coming to consciousness of the presence and potency of the elements, connecting with the physical and spiritual sustenance provided by those elements, and cultivating a set of practices for living in a way that is deeply influenced by the elements.

First, one group of authors focuses more on the elemental life as a life that seeks a humble, ever-evolving awareness of the elements and their powers to create, shape, destroy, and heal. Liz Beachy Gómez highlights the Brazilian tradition of Umbanda, as well as rituals that remind us of how our inner and outer landscapes mirror each other and connect the elements with people's sense of a larger self. Gomez reminds us that "each of us carries our own exquisitely unique elemental composition. Some of us are predominantly Fire, while others carry more Earth, Water, or Air." David George Haskell rethinks how we might redesign the periodic table of elements, calling for scientific iconography that emphasizes "nonatomistic truths" by better illustrating the interconnectivity of the elements. Gavin Van Horn revisits Gary Snyder's imperative to "go light" as a challenge to "join our consciousness—our inner light—with the greater luminosity of Life as it is expressed through our fellow earthlings." Vandana Shiva calls *akash* (which she translates as "space") the fifth element, claiming that space is "the organizing principle of the elements" that allows for relationships among all elements to flow dynamically.

Second, another group of authors emphasizes ways we can more effectively connect with the elements. Leeanna Torres speaks to the elements of adobe in her New Mexican community, detailing adobe as "the map of our story, the materiality of our daily lives, earth and water, water and earth." Yakuta Poonawalla talks about the power of her family's metropolitan terrace in Pune, India, where she can best "understand, sense, and feel the connection between us humans, the natural world, and the elements around and within us." Suzanne Kelly speaks to the elemental challenges and interactive power of mud as a garlic farmer, conveying that "mud can be chaos," but the kind of chaos that reminds us of our vulnerability in a way that deepens the possibility of "connection for the living in a time of gaping absence of relationship." Sophie Strand adapted a selection from her memoir *The Body Is a Doorway*, sharing: "I think my mind is not just in my body. It is... my entire web of relations—fungal, geological, microbial, vegetal, ancestral—that weave together my specific ecosystem.... I imagine that I am like a mycelial network below ground, opening up the septa pores in my branching hyphae." In my essay, I talk about how care is elemental, reflecting on how individuals, a small community, and the elements themselves began to heal together through a climate-chaotic Vermont flood.

Closing out the volume, a group of authors outlines specific practices for living more elementally. David Macauley observes how the elements are "deeply present in many forms of walking, especially when we stretch our environmental imaginations." Mark Coleman shares with Van Horn his "elements meditation," explaining that meditation can deconstruct "that we are a separate self" which helps us "feel less alienated, less separate, and we start to feel that we are part of a web that we are affected by and have an impact on." Joerg Rieger looks at the ways socially just human labor can put us in conversation "with the elemental nature of other-than-human labor and work, as performed by earth, wind, fire, and water." Carina Lyall shares her own intentional labor by describing herself stroking a fire striker to make a fire, concluding that such practices enable us to become "apprentices" of the elements. We purposefully end with Priyanka Kumar, because her essay on play reminds us that the elemental life should inspire a "state of mind" that suspends our sense of time, allowing us to be more present with and absorbed in our elemental connections.

So, let us play.

Elemental Space: An Interview with Vandana Shiva

John Hausdoerffer

John Hausdoerffer: I am excited to see that your autobiography, *Terra Viva*, is out. Congratulations! I love the title, so I am curious about it. Would it have been fair to call your biography the same thing as this volume, *An Elemental Life*?

Vandana Shiva [*laughs*]: Well, my publisher picked the title from a manifesto I had done on connecting soil, climate change, and the refugee crisis. For me, my life has been about the living Earth, working with and defending the living Earth. So, *An Elemental Life*? Yes, because I do defend the elements and I feel for the elements. In science, we always had a phrase, "feeling for the organism." The reduction of the elements to separate entities—rather than a kinship relation—has always been an epistemic separation. If the elements are separate, and if we are not also part of the elements, then how to know them, feel them, and defend them as kin becomes an impossible task. That is why Western philosophy has never been able to figure out the notion that if we are elemental, then the knowing is relating. So, yes, *An Elemental Life* could have been a possible title.

John Hausdoerffer: Yes, the kinship relation is vital to knowing, feeling, and defending. Our last project, *Kinship: Belonging in a World of Relations*, was a five-volume project exploring the kinship that exists between human and more-than-human beings (with the five volumes scaling from persons to partners to place to planet and then circling back to practice). This new five-volume project, *Elementals*, looks at the classic elements through a kinship lens. And this fifth volume is called *An Elemental Life*. What does it mean to live a good life when seen from our place in the kinship networks formed by earth, air, water, and fire...

Vandana Shiva: And space! In the Indian cosmology, space is the fifth element.

John Hausdoerffer: Oh, wonderful! Actually, one of the major questions I have for you is how you weave together ways of knowing and cosmology as a physicist, as a Hindu woman, as an eco-feminist philosopher, and as an activist with an eco-centric worldview. We have found that "the elements" over the course of the history of the West have been atomized and overcategorized, separating humans from the kinship networks that are truly our home. From your point of view, what are the elements?

Vandana Shiva: The elements are not the reductionist chemical elements with which the periodic table makes us familiar. Instead, the elements are the constitutive elements that, in interaction with each other, connect humans to all of the earth, nature, and the universe. So the elements are the basis. *Basis* not in terms of broken constituents but in terms of dynamic relation, the very constitution, of relationships.

In Indian cosmology, as I said, the fifth element—space—is the most important. It was falsely interpreted as ether in Western philosophy and science, and then bashed.

Akash is space. Unlike air, we cannot feel or measure space, but it is the organizing principle of the elements. And so, of course, for Einstein, the interconnectivity of space and time was the contemporary version of it. But in Indian thinking, it was understood to be place-as-space (which is relationship) and God (which is time). Place and God, space and time, are not separable. So not only was the periodic table flawed in atomizing the elements; space and time were separated. Time was made linear and space was made to disappear.

It is extremely clear that elements and living systems are dynamic, interactive elements. Water and earth and air and fire are constantly interacting in the earth's body and in our bodies. And the balance, the right balance among them, is what makes for health. Healthy ecosystems and human health. I mean in Ayurveda, the principles of Ayurveda are balance between the *doshas*, our energies, you know? Doshas that are the expressions of the elements.

When I got involved with biodiversity movements in the 1970s, I was compelled to look at what happened to agriculture in the context of the violent impacts of biotechnology and genetic engineering on farmers, the supposed "second green revolution." I played a very big role in introducing Article 19.3 to the Convention of Biological Diversity at the 1992 Earth Summit in Rio de Janeiro, which was about protecting life on earth and protecting biodiversity. Article 19.3 was about the impact of what they called GMOs, LMOs (living modified organisms), on biodiversity.

I talked to government after government in the lead-up to Rio, saying, "If you're looking at biodiversity, then we must look at the impacts of these technologies on biodiversity." So we managed to get that clause, 19.3, into the treaty, and then, because I was the only one talking about GMOs at that time, up to the Earth Summit, the United Nations invited me to join an expert group to draft the framework of a protocol that would implement 19.3. I won't go into the details of that expert group. It has its own stories. But we managed to put 19.3 into the treaty and President George H. W. Bush walked out on signing the Convention of Biodiversity at Rio, saying he would not allow this fifty-billion-dollar industry to be undermined by regulations of any kind. And when he went back, his vice president, Dan Quayle, issued an order saying that from then on, a genetically modified organism will be treated as if it were not modified. The Bush administration announced the term *substantial equivalence*. This is a doubly funny story because

Quayle's the one who once visited a school and couldn't spell *potato*, you remember? He couldn't spell *potato*, but he could spell a very complex word like substantial equivalence.

Now, substantial equivalence is not just false equivalence, because a genetically engineered organism is not equal to a non-genetically-engineered organism. But second, legally, the same companies that say it is the same when it comes to avoiding responsibility claim novelty when it comes to taking property rights.

John Hausdoerffer: Right, so Quayle's and Bush's victory for the industry was that if a GMO was "substantially equivalent" to a regular crop, then there was no need to label or test that GMO. My readers might be asking how this moment in history is so important to a book that rethinks the elements. Is this why? The capacity to claim substantial equivalence between genetically modified and organically evolved life *relies on* a previous, long-standing, and deeply accepted reductionism of the elements, rooted in atomizing and categorizing them in the periodic table. That is, because we can reduce and separate those elements so easily, it is thus much easier to accept substantial equivalence. But if we saw all elements as kin, evolving together *in a fifth element of space*—with humans as cocreators—in a shared process, it would be harder to reduce the elements to separate entities and, in turn, to segregate our elemental kin as resources available for exploitation and equivalence. Is this a fair summary of the significance of substantial equivalence to a conversation seeking to reconnect the elements?

Vandana Shiva: Yes. After substantial equivalence was widely accepted, all the discussions then became about chemical constitution. You then had ridiculous statements like "Oh, we've done the biosafety," or "We've genetically engineered buffalo and it doesn't have a different weight from a non-genetically-engineered buffalo," you know? Or, "The chemical constituents are the same," but the process of it had totally changed. This to me became problematic not just in terms of genetic engineering but also because England had started to feed its cows and sheep dead cows and sheep. And of course, that was totally justified as an example of substantial equivalence to the diet. Equivalent calcium, equivalent iron, equivalent everything, but these were dead animals. Then the mad cow disease hit. The mad cow disease, of course, led to a transmission to humans, and earlier they had insisted it could not shift to humans. A scientist earned a Nobel Prize for deciphering what had happened, finding that it wasn't an external infection and it wasn't a virus or a bacteria; it was a self-infection.

Here is where space as a vital fifth element comes in. The self-infection came from a deformed protein. It was the equivalent protein in terms of its chemical constituents, but it was distorted in space, and just that certain shift in space had made it into a self-infected agent. Again, this is why space becomes extremely important. Space linked to time becomes even more important. Space and time are a process, because nothing is static in the world. In the living world, everything is dynamism. Reductionist, atomistic, rigid understanding of the elements allows you to skip the most important aspects of life, which are process and dynamic change and harmony and balance.

John Hausdoerffer: It seems that the fifth element of space is incredibly significant in terms of worldview. Is the space within which the elements move together a community of human and more-than-human kin? Or is the space within which the elements rest a global stockroom of resources awaiting exploitation? These are two very different frameworks of space leading to two very different relationships with the elements and two very different ethical ways of being, two different views of an elemental life. I am thinking of some of your classic publications. Is a kinship relationship with the elements less about how we view Earth—like in your book *Soil Not Oil*? Or how we view water—like in your book *Water Wars*? What I think I'm hearing from you is that the relational view of elements will follow if we revolutionize our view of space. It seems like space is the catalyst for a less reductive view of the elements. Rethinking how we view soil or water or anything will follow a rethinking of space as process. Is that right?

Vandana Shiva: Absolutely. What's coming to my mind as you're talking is the Chinese philosopher Lao-tzu. I remember a very powerful quote of his. He says you can carve out a window, and it's the space within the window that makes it significant. And you can throw a pot, but it's not the pot; it's the space within that makes the water-holding capacity of the pot. So space is vital. Space is where relationships occur.

In between elements is where relationships occur and changes in relationships also unfold in space and time. Otherwise, we will mistakenly see each element as separate from other elements, leading to the totally mechanistic understanding of a separated universe. Even worse, this separate universe will also seem static because it is the unfolding in space and in the relationship that you get the potential for dynamic change and metamorphosis.

John Hausdoerffer: This reminds me of the metamorphosis of farmers' struggles that you imagine at your farm Navdanya in Dehradun, India. You empower farmers to build a new resiliency in the face of climate chaos (*Navdanya* means "nine seeds," to reflect the diverse, adaptable food systems needed so farmers always have livelihood no matter the weather at the time). Does your work at Navdanya represent a return to a certain view of space? It is certainly an innovative return to and advancement of certain agricultural practices that allow for people to stay on the land; cultivate a life of dignity; produce as much per hectare as under pesticides and fossil fuels; allow for independence from biopiracy, independence from the domination of the fossil-fuel and agribusiness industries. That was all clear from my time there with you, especially in the face of dramatic farmer suicides across India, suicides that can come from reliance on systemic underdevelopment and poverty. Navdanya restores so much in terms of both spiritual and substantive hope—hope for everything from political efficacy to personal dignity to soil health to climate resilience. But I'm curious how the work of Navdanya restores this renewed view of space for the farmers who participate.

Vandana Shiva: Well, three ways. First, in the dominant industrial paradigm, farmers are supposed to disappear. I remember, early in my time working on agricultural justice, I wrote a report on agriculture in which I tried to make sense of why agriculture in the United States still made America an agricultural power. Then, upon reading Wendell Berry, I realized that agricultural dominance in America ironically comes through making farmers disappear. It wasn't a revolution for the farmers. I remember a quote from an agriculture secretary in the early 1990s. The secretary said you have to squeeze the last farmer off the land—which is *bhoomi*, which is the earth, which is one of the elements—like we squeeze toothpaste out of a toothpaste tube. So the land is reduced to the metaphor of the toothpaste tube, and farmers are reduced to the toothpaste that must be emptied out. And the ultimate in all of this—Mr. Bill Gates and all his followers in the world today have this dystopic vision of farming without farmers and food without farms. This dystopian vision is built on a separation from the elements.

John Hausdoerffer: And the view of space. The dominant view of the space of the farm will become a space without the farmer.

Vandana Shiva: Exactly! All of my work—primarily my research work before I started to work in seeds and created Navdanya—was always about ecosystems and ecology and ecological processes, and I always, in my assessment, had the concept of ecological space. That a forest has a right to its ecological space. A river has a right to its ecological space. Ecological justice for human beings is having a right to their ecological space. You could take the right to clean air and clean water as a right to ecological space. Freedom from toxics as a right to ecological space.

The farm in Navdanya taught me things, oh my gosh. First, that there is so little land left, and now these days, my preoccupation is dealing with the land mafia, a whole new experience for which I am prepared, but I am having to learn how to deal with the worst. I have had to deal with the cartel and the mafia. But our relationship with farmers requires (a) shifting the space away from the view that they must disappear and be part of the extinction crisis and (b) helping them see that they have a central role in shaping the future with the right balance and the right care for the elements. Increasingly, I see agriculture as caring for the earth. Not as a productive activity. Production is a by-product of care. The farm must first be a caring space.

John Hausdoerffer: And the farm must first be a caring space, for and from the farmer as a caring being, before it can care for the health of the elements necessary for farming in an age of eroding soil (earth), floods and droughts (water and air), and drier and extended burning seasons (fire). I suspect I am seeing only the tip of the iceberg, but I am starting to understand why you see space as the fifth and vital element.

Vandana Shiva: Last year, there was terrible heat in India. There was also extreme rain. But during the heat, and it was so very hot, the crops on our farm were doing fine. The crops wilted on a few neighboring farms. I purchased a hydrometer—a small little gadget to measure moisture and soil temperature. I have never accepted the monoculture of the mind that reduces agriculture to row crops grown as commodities. Agriculture is about right relationships between diverse species, and therefore agroforestry is part of agroecology, even though my own colleagues in the early days said, "No, no, no, don't plant trees around the field." I said: "But all farms have these field trees. All our ancient farms have farm trees around the field." I was stubborn and continued to plant. So this summer we measured between our field where we have trees and the neighboring farms. The difference in temperature was twenty-five degrees centigrade. And that's just the soil temperature.

John Hausdoerffer: Seeing the farm as a space of agroecological care is what has made the difference in climate resilience for the soil and for the farmer?

Vandana Shiva: Yes. This also tells you the earth, the water in it, the fire, the temperature in it and in the air, are all interconnected! That's why the whole kinship relationship among the elements is so vital. The moisture difference was 15 percent. Our philosophy is that there is no element that you can get rid of; there is no species that is disposable. You must maximize biodiversity, and to maximize biodiversity you need care and you need to care for caring farmers. Therefore, instead of farmers being disposable, farmers are central. The space shifts, you know? There is space in farming. Instead of being on the margins to be pushed out, it becomes key in the relationship in healing the regeneration of the right relationships.

John Hausdoerffer: There is space in farming, and it seems that space—for you—is one that centers the farmer as a cocreator of an agroecological space. We know that—in a true kinship relationship—humans and all other organizations of the elements are in the same social-ecological space as one family. Are you saying that the farmer is the beginning of space itself in your ecological vision, in an age of climate chaos?

Vandana Shiva: Yes. Yes. We have a very strange Western interpretation of history, because too many people have just assumed that ever since people left the forest and started to farm, we've had problems. But if that were the case, how would aboriginal people in Australia farm for over ten thousand years? How could the indigenous people of the Americas have farmed for thousands of years? How could the Indian farmers have farmed for five thousand to ten thousand years? Or in China for forty centuries? The fact that farming has sustained itself over those periods of time means that farmers are key to creating space for all species, creating relationships of kinship, and through those relationships of kinship, creating the conditions of renewal and regeneration. Humans as a species can find their space but also enlarge the space for others. That is what it means to be a cocreator.

The Navdanya farm is a good example. We are next to the forest, intentionally, but the forests are getting so degraded. The animals are not finding food, so we have to take care that they do not destroy the seeds that we are saving. There is literally a job at night for one of our colleagues to walk with the phone's flashlight to scare off animals. Once we had twelve elephants. The farm wasn't an elephant land, but it is now because there is no longer enough space in the forest for wildlife like elephants. That is why my life's work and research has shown that the compromising of the fifth element of space from the maldevelopment process is at the root of conflict.

John Hausdoerffer: You really have me thinking of space as a fifth element from a systems perspective. The work you are doing in finding those *navdanya*—those nine seeds that allow for women to lead in climate adaptive restoration—means that your farmers are living an elemental life in a way that cocreates the resilient future of the elements themselves. They are creating a space that reduces soil erosion (earth), that adapts to declining Himalayan glaciers and stream flows (water), that adapts to a warming climate (air), and that plans ahead for a world of increasingly more intense burning seasons in forests and fields (fire).

In addition to the systems understanding of the elements, it also strikes me that kinship as an elemental life practice emerges from this system you have given your life to cocreating, democratically, in your region of Dehradun. For example, you have focused on re-enlivening the practices of women as a knowledge-generating, bio-diversity-enriching keystone species for at least eighty thousand crop species in agricultural history. There is something deeply elemental here. I am wondering what this looks like as a way of life. I am thinking of when Henry David Thoreau said he went to the woods to learn to live deliberately, and his time in the woods gave him the clarity to protest slavery through civil disobedience, leading to a night in the Concord jail. Or I think of Socrates, who said that the unexamined life was not worth living, knowing that pushing the Athenian jury to examine their lives would lead to his own execution. For you, if we replaced some of these statements with your own, what is the elemental life?

Vandana Shiva: All of the elements are at play all the time. When we understand space as the fifth element, everything else gets expressed within it in different relationships. I agree with Lao-tzu on the lessons of water. He highlights the paradox of water carving rock into deep canyons—what is fluid will overcome, drop by drop, what is unyielding. What is soft and adaptive will overcome what is hard and rigid. There is a feminist song with the line "drop by drop." Drop by drop we wake up, with gentleness and fluidity. That is why eco-feminist philosophy really shows the power of nonviolence. It brings together both sides of Lao-tzu—through gentle nonviolence we are fluid, yet we are also the rock, standing strong in our relationships. The power of nonviolence reveals the power of the elements being in balance within a space. That is what it means to be elemental, to be in relational balance within a space. It means having the fluidity and strength to face the harshest, cruelest, most brutal power.

John Hausdoerffer: What does that look like in daily life, for someone not involved in a nonviolent action like the 1970s Chipko movement in Uttarakhand, to which you have drawn global attention, in which women hugged trees to protest logging's threats to their livelihoods and the hundreds of lives lost as a result of flooding from deforestation? What does an elemental life of fluidity and strength—of being the water and the rock—look like for different kinds of readers in different kinds of regions engaged in different kinds of activities? What might be universal about an elemental life for people to try on for size?

Vandana Shiva: First is to understand that Chipko (and many like it) was not a movement, and that's the thing about the elemental life—its always a process. It is always a process. It took a decade to decide. It took a decade for the women's message to be heard. When floods and other disasters kept increasing from deforestation, the women were saying, "This is the real value of the forests— the standing forest holds the value of a forest, how it provides clean water and air." Can you see how powerful the elemental understanding is? It is only when the floods became worse and flood relief became bigger that the government woke up and noted the "ecological services" of the forest, even though for the women their elemental philosophy had told them that for centuries.

I learned just today about how this process continues. On December 2, 1984, a leaking Union Carbide pesticides plant killed thousands. The women are still protesting. Ten of them announced today a hunger strike without water. People are dying to get justice, even today. Ten women, willing to give their lives for the lives of the community. That's elemental. Because the kinship piece is not just about me; it is about life as a collective of past generations and future generations coming together in space and time. The two women who have begun this remembrance protest in Bhopal—one was a Hindu and one was a Muslim. These women hung together for justice, in the midst of constant attempts to divide Hindus and Muslims. Now these ten women are saying, "We are here."

John Hausdoerffer: That's kinship. That's elemental kinship.

Vandana Shiva: Yes. I used to go to every anniversary. I remember, on the third year, one of the women said: "Until we get justice, the fire's an element. The fires that cremated our lost ones, we'll keep burning in our hearts."

This interview took place on Zoom on December 30, 2022. Special thanks to Gillian Bauer, for her work on the transcription.

Atoms on the Wall: Elements and Relationship in Science Education

David George Haskell

I drop the stack of graded lab reports on the front desk of the lecture hall, then turn to poke a button on the wall. A motor in the ceiling whines and a wide metal bar descends, pulling down a projection screen. I will show no slides in class today, though. Instead, I'm using the screen to veil a huge periodic table of the elements. Six feet high and about twelve feet across, blazoned on the wall above the teacher's podium, the periodic table is the only lettering or artwork in the classroom. In this traditional lecture hall, all the student seats are bolted facing forward. And so, regardless of the topic in their class, students must stare at a gigantic representation of elemental life.

The table represents a singular authority. It hangs in the same location as portraits of a "dear leader" or "royal highness" in classrooms in dictatorships and monarchies. Like a portrait of a despot, the periodic table on science classroom walls displays one truth. *Gaze on me and learn, young minds!* But there are many truths in science. Elemental facts and processes exist at many different scales, from subatomic to ecological to cosmic. The periodic table focuses our gaze—quite literally in many classrooms—on just one time and physical scale. As a teacher, I want to wander with my students through many times and places. It is hard to do so under a propagandist banner.

The periodic table purports to show the elements that constitute our world. It makes a strong claim: *Here is the foundation of the universe.* As such, the table is also an ethical manifesto. Knowing the "nature" of reality, we can then decide who we are and how we should act.

Note two prominent features of the table. It is atomized, and it is ledgerlike. The atomism of the table teaches us that the world is founded on separation and individuality. At root, it says that everything is alone and belongs primarily to itself. Any relationship that emerges from this selfhood is secondary to the atom. Then, the ledger. By using ranked and numbered rows and columns, the table states that the elemental form of the world is that of the accountant's spreadsheet. Numbers and weights. Aligned rows. If the world is a ledger, then to buy, sell, and otherwise objectify and trade its riches is entirely "natural."

The table, though, is not as elemental as it appears. Its authority and order are partly illusory. Yet before considering the table's limitations, let me note, as I do for my students, the alluring and marvelous beauty of the story told by the periodic table. Taken at face value, the table tells us that the varied substances of our everyday world are all made from a few seemingly immutable particles, the atoms, also called "elements" in most modern chemistry classrooms, displacing more ancient ideas of elements. Among living beings, about a dozen of these atoms suffice to build most of the animate world in all its diversity. The aromas steaming from your cup of coffee, the wood supporting an oak tree, and the spiderweb's silk are all made from just five: carbon, hydrogen, nitrogen, oxygen, and sulfur. Add a few more to this mix, and you have the atomic foundations of all of life on Earth. These elements are surely the universe's most expressive alphabetic language. They speak the world as we encounter it into existence.

Further, the table suggests that these atomic particles are not a jumble, like letters in a bag of Scrabble pieces. If we arrange the elements by weight, a repeating pattern emerges, a periodicity like the octaves in music. This was the great insight of the nineteenth-century chemists John Newlands and Dmitri Mendeleev as they developed the first versions of the table. If we know the position of any atom in the table, we can predict its properties, such as whether it is a reactive metal or an inert gas. We've hoisted this primal spreadsheet on the wall, an icon representing the order of the world.

But take another look. Much of the table's order emerges not from the properties of the elements themselves but from how we've chosen to arrange them. In the first three rows, the columns line up because we've inserted huge gaps into the rows. The sixth and seventh row line up vertically only if we excise fifteen atoms from each. Some elements defy the supposed unifying properties of their columns. Hydrogen is not a metal, for example, and mercury is not a solid. The periodic table reflects a human desire for tidiness as much as the fundamental properties of atomic elements themselves. We've massaged the data to create a ledgerlike image. When we look at the periodic table, we're gazing not at elemental truths about an objectively observed world but at a mirror into our own psyche and culture.

The periodic table focuses on just one scale of reality. Although chemists in the 1800s could not know it, their tables pointed at particles and energies even smaller than the atoms on their charts. The "elements" are not elementary. An "atom" comprises a tight cluster of protons surrounded by electrons. For each atom, the numbers of protons and electrons are always matched. Nitrogen has seven protons and seven electrons. Oxygen has eight of each. The electrons arrange themselves in "shells" around the proton core. The inner shell can fit two electrons, never more. The next shell fits eight. The next, eighteen. The nested organization of electrons explains some of the regularity in the periodic table. The chemical properties of an atom depend on the fullness and arrangement of its electron shells. Atoms with full outer shells are inert. Those with partly filled shells are reactive. For example, hydrogen has only one electron in its inner shell. Add a whiff of oxygen and a spark: boom! Hydrogen's half-empty shell makes it

explosive. Helium, though, has a full inner shell with two elec-
trons. It is inert, even when provoked with reactive atoms or heat.
Electrons, not "atoms," are running the show.

Perhaps we should post on the wall a visualization of electrons
instead of the periodic table. This would reveal a different elemen-
tal life. Instead of a table proclaiming that reality is atomized, a
depiction of electrons would emphasize relationship. Even when
physically separated, electrons can remain united to others, some-
how "knowing" what other electrons are doing. What might this
"spooky" quantum entanglement teach us if we painted it on every
science classroom wall? Perhaps that all matter and energy are
stitched together in mysterious ways. Connection, not separation.
The ethical lesson would be very different. If we're forever and un-
alterably connected to one another, then belonging, not aloneness,
is the foundation of the universe.

As a biologist, I find the neat rows and columns of the peri-
odic table especially grating. Life is not tabular. Life does not fit
into tidy, predictable patterns like those of atoms or even electron
shells. The evolutionary tree of life, for example, is wildly lopsid-
ed. Some branches, like beetles and flowering plants, splay out in
profusions of hundreds of thousands of species. Others, like tua-
tara and cycads, are singular twigs or close-cropped clusters of a
handful of species. What is true in genealogy also applies to phys-
iology, microbiology, genetics, and ecology. This does not mean
that life is a random jumble with no unity. Rather, the elemental
forces of life—genetic inheritance, evolution, ecological relation-
ship, energy pathways through ecosystems—play out in thrillingly
diverse ways, nudged by the happenstance of history and life's
creative processes. Nothing is atomized. Everything is relational.
Thus, to teach under a placard made of rows and columns seems
antithetical to my understanding of life's processes.

Hiding the periodic table behind a screen is hardly a complete
answer to this problem, though. Discussing the strengths and weak-
nesses of the table must accompany the veiling, a conversation that

hopefully brings to the surface the ways that all scientific diagrams carry hidden meanings. We then read the periodic table for its information but also for the assumptions and values built into its iconography. I find it helpful to ask similar questions about biological diagrams. Evolutionary trees, for example, show the geological relationships among species. The treelike picture, though, assumes that creatures inherit their DNA from their parents. But we now know that many living beings also pick up DNA from neighbors. In such cases of "horizontal gene transfer," the metaphor of a tree is misleading. A braided river delta would be a better image, so the "swampy river of life," not the "tree of life." Evolutionary trees also imply that the identity of a species is primarily determined by genetic ancestry. Blood ties. You are what your family was. But another view is that ancestry is less important than ecology and local context. The same species might have very different roles in different ecosystems and thus more than one identity. In this view, genealogy is less important than relationships with other beings in the present moment. The point is not that the periodic table or an evolutionary tree is wrong. Rather, every diagram is a simplification of a complex world. Each one is a storytelling device, and stories often have concealed or subconscious layers of meaning.

What might a biologist display at the front of the classroom? There can be no single answer to this question. Life is too diverse and unfolds at too many scales of time and space. The most straightforward answer, then, is an ever-changing sequence of images representing different timescales and processes. How about a "periodic table" from the first moments of the cosmos' birth? Atoms had not yet formed, and the universe was a roiling, pressurized soup of protons, photons, and electrons. This reminds us that the "elements" look different at different times in the story of the cosmos. Then, let's post images of the hundreds of species of bacteria, fungi, and tiny animals that live inside a tree leaf, a reminder that the most everyday living beings are not individuals but living communities.

For every day, a different image. Stories unfold from each one. Instead of presenting a single idea about the nature of the elements, we'd show that elemental life is multifarious. If the artworks and diagrams in our classrooms are drawn from multiple cultures and time periods, we could also learn that all images carry within them the values and priorities of their creators. Images of elemental life, then, bring us into conversation across cultures and across time, and they may also reveal the diverse ways that ethics and meaning are understood and represented in art and science.

What of the classical elements in classroom art? Surely few modern science classrooms depict the ancient Greek elements, fire, earth, water, and air (and the fifth heavenly element, aether). Likewise, we do not post on the wall the classical Chinese elements, wood, earth, water, fire, and metal. Yet I think these images belong alongside more contemporary images in our classrooms. They remind us that curiosity about the elements has deep roots across many cultures. We've been searching for elemental roots for millennia, and as the mysteries of quantum physics show us, we have not completed the quest. Modern images, such as the periodic table, are mostly nineteenth- and twentieth-century attempts, likely to be superseded as cultural values evolve and knowledge deepens.

The classical elements also contain insights missing from modern iconography. They embody the idea that elements exist only in relation to others. The indivisibility of elements (from the Latin *elementum*, "matter in its most basic form") implies an essential individuality and separation—literally, atomism (from the Greek *a-*, "not," and *tomos*, "cuttable"). Yet in the classical conception of the elements, individuality depends on connection. This contrasts with the periodic table, where the existence of one atom in no way depends on any other. Carbon does not need lithium. Each classical element, though, exists only in dynamic relationship to the others. Without any one of the elements, the physical world would be nonsensical and the cosmos out of order. In the

Chinese tradition especially, the relationships among the elements are as important as the elements themselves. Elements generate one another (wood begets fire, for example) and also suppress one another (water quenches fire). In this cosmology, the patterns of these connections determine not only the physical properties of the world but also human physiology, political processes, and the order of the universe.

The classical elements therefore show us how to live inside a paradox. Complete individuality and total interconnectedness can both be true at the same time. This is a refreshing antidote to the stories of modern science and economics that, until recently, emphasized individuality rather than relationship. The atomism of the periodic table is also present in the foundational ideas of evolutionary biology, genetics, and economics. Charles Darwin and Adam Smith both theorized that the history and future of life and the human economy emerge from individuals interacting and pursuing their individual goals. What are the fundamental, elemental units in those theories? Individual organisms, genes, and economic actors. These approaches yield wonderful insights. But they are also limiting. We now know that a gene cannot be understood separate from its interactions with other genes. In fact, a gene is just one node in a giant web of conversation within the genome (and beyond, with symbiotic microbes, for example). An individual organism is a living community of dozens or hundreds of species (thank you, gut and skin bacteria, for being essential parts of each one of us). A species is a convergence of multiple evolutionary genealogies (all hail the mitochondria in every cell of our bodies!). The classical elements show us that extreme individuality and deep interconnection can both simultaneously be true. Individual atoms, molecules, genes, organisms, and species are not just in relationship; they are *made of relationship*, generated, animated, and sustained by interconnection.

Being made of relationship does not erase individuation. Rather, the convergence of relationships generates individuality.

The nature of this convergence gives each "self" its character. A tree, for example, is a living community whose vitality is especially dependent on links with fungi and bacteria in leaves and at root tips. These networked relationships allow the individual trunk and roots to explore the soil and lift the tree community to the sun. Human individuality emerges both from biological relationships, especially with microbes in and on our bodies, and from the particular psychological and cultural relationships in which we grow. The nature of individuation, then, is a property of the network. This is true for chemical atoms, too. Interconnections among protons and electrons produce substances, atoms, each with distinct properties that emerge from the relationships among subatomic matter and energy. We live in a society where individuality is assumed to be supreme. I have a name, a Social Security number, and a CV, and one day, I will have a gravestone. Each one of these fundamental cultural properties speaks of individuality. Yet the elements remind us that relationship and interconnection are just as important. What if our governmental and business labels, monikers, professional documents, and memorials emerged as much from relationship as atomism? When we are embedded in highly atomized cultural structures, it is hard to imagine such a world. Some inspiration from other, less atomized cultures might help. For the Waorani of the western Amazon, for example, the names of people and of plants arise from context. Your name emerges from your social group, and if you move, your name changes. A plant "species" is known by different names depending on where it grows. In the modern Western world, though, name changes for humans happen mostly only in the context of patriarchal traditions of name changes after marriage. In contemporary taxonomy, nonhuman names such as Linnaean epithets do not depend on ecological context.

The periodic table of the elements, then, reinforces the societal narrative that the "self" is supreme. Little wonder, then, that we take such poor care of one another and other species. If lithium is irrelevant to the nature and existence of sodium, why should sodium attend to its relationships with that other atom?

All teaching involves a hidden curriculum, the assumptions and values that we convey, often without conscious awareness. Images on classrooms walls are part of this unspoken indoctrination. But the images can also be fun and thought-provoking starting points for conversations and lessons. Education is at its best when the hidden is brought to the surface and examined.

In my teaching, reimagined classroom icons have been only part of the answer to uncovering the "elements" of the world with my students. My experience teaching under the giant periodic table spurred me to metaphorically unbolt the seats in the room. Lowering the screen is an insufficient answer to atomism. Instead of facing the wall, we face outward. No wrenches are involved. I make sure that at least half of our class time is scheduled outside. There, students are fronted not by a wall, diagram, or podium but by the weather, passing birds, campus plantings, vehicles, and people on the move—the relational elements of everyday life. Every class has an objective, of course, but getting outside shifts the agency away from me and my teacherly planning to the unplanned possibilities of the moment.

Earlier this year, for example, crows redirected the agenda for my writing class. A pair were rattling quietly to one another in the high branches of a tree behind a campus parking lot. Our focused study of human language then expanded into the varied languages of other species. Creative-writing students learned that crows don't just "caw" but have complex family lives mediated by nuanced and sometimes mysterious vocalizations, and that listening to what seems peripheral—birds were not part of the syllabus—can be more interesting and generative than what we "should" be focused on. I learned from this that giving up control of the flow of the class to relationships with other species induces both anxiety (I'm not meeting "learning goals" about writing craft; the students might skewer me on their evaluations) and delight (thank you, crows, for breaking us out of our bullshit little boxes—now we're really doing "creative" writing). A few weeks later, the campus was strewn with pellets of lawn fertilizer. My environmental studies class considered the sources and the destination of the granules, putting our everyday experience of lush campus lawns into relationship with upstream phosphate mines and downstream rivers. We also pondered the origins of the putting-green aesthetic and asked who bears the costs of our desire for close-cropped grass. Again, I was thrilled that our class was exploring the strands of interconnection that surround us but also worried about "covering" the promised class material. Of course, impromptu and embodied learning is far more memorable for students than any typed-up syllabus, but the tyranny of the rows and columns of class "assessment" sheets is deeply ingrained. Covering up a periodic table is easy to do. Unlearning decades of external numerical evaluation and ensuing self-judgment as a teacher is not.

If elemental life is about relationship, then our classes need to break out of hermitic classroom walls that block nearly all possibility of embodied connection to the world. It is not enough to put atomized brains in a room. We must also engage our whole beings in the elemental relationships from which we are made.

In the Time of Mud

Suzanne Kelly

I heard the warning loud and clear in the Canada geese's clamorous sky honking. I even stopped to get a better look at their formation and direction, laying down my bucket of garlic seed to shield my eyes from the sun. But with their adequate food supply and dearth of predators, with carbon dioxide warming their air and water, their calls had taken on a dubious quality in recent years, as many of them had become year-round residents, continuing their stay at Mill Pond—just across the hayfield from my farm—throughout the winter. Skeptical the seasons would change at all, I solemnly carried on, wanting to stay on schedule.

In the northeastern United States, where I live, garlic is sown in the fall. Planted too late, the alliums won't have enough time to establish a strong root system to get up and go with the spring thaw. Too early, and they may sprout during that charmed finale of a warm spell, which doesn't usually mean the end of them, just unnecessary human worry. Because I've been known to worry, I've typically erred on the later side for sowing, with a goal of the third week of October. I've seen images of industrial, big-ag planters in China that can get twenty-five thousand garlic seeds in the ground in under an hour. But most farmers throughout the United States, including myself, continue to sow the heads by hand, one clove at a time. I once grew as many as twenty-five thousand, but with the addition of more and more crops, I've scaled back to about half as much. That nonetheless amounts to nearly a week's worth of labor, if you include bed prep and mulching, at a time of year when only the oaks and silver maples are hanging onto their leaves and the

cruciferous crops are hanging out in the field; when most every-
thing, except maybe the aberrant saffron crocus, is either descend-
ing downward or drying up. When time is running out.

I'd planted eleven of the fifteen hundred-foot beds during one
of the wettest Octobers on record, with a few of the paths three
inches deep in water on land I'd never seen flooded before. Despite
my rubber boots and a rain slicker, squatting for hours left my feet
and hindquarters tingling and so senseless that by the eleventh
bed I packed it in for a hot bath.

Four beds to go and soil still sodden, more stoppages emerged.
My partner suddenly fell ill and was hospitalized. And then we got
five inches of snow.

A week later, my partner home and on the mend, I ventured
out to the field to see what, if anything, was possible. Most of the
snow had melted, but the temperature had also dropped again,
freezing the thin upper portion of the puddled water into something
like sugar glass that crackled when I stepped on it. Later that day, I
ran into a fellow farmer who nervously asked me if I'd gotten all my
garlic in. "Not yet," I shrugged, explaining my derailments. "But the
weather looks mild enough in the next week that I expect to finish
up." After all, I thought, it's only the first week of November. There's
still *time*. A fourth-generation fruit farmer with Canadian roots, she
appeared unconvinced. "When it snows on mud," she said, "it's over."

That afternoon, temperatures rose into the high forties. I
bounced back, too. All the snow and ice had disappeared, leav-
ing several inches of soupy loam that was sometimes difficult to
contain inside the edges of my thirty-inch beds. It was sloppy, but
the seed was ready to go. With my boots earth caked and my legs
fatigued from carrying them, I planted the last of the cloves.

What I love most about growing garlic, after the pleasures of cook-
ing with the bulbs and eating them, is its early emergence and

tenaciousness in reaching for the light. Unlike other early risers, like the bell-shaped snowdrop and hellebore that are also known to break through frost and snow but at full bloom bend away from the firmament, garlic's inner stalk pushes from seed to flower with a consistent verticality. As soon as the sun begins to move closer to the earth, the garlic will find a way to grow toward it. It was no different that following March.

The returned proximity of the sun also brought back the muck, in the garlic beds but also everywhere, as the ground defrosted and mud season, as it's sourly called, ensued. It was especially insidious down the road at our town's green burial ground, where, when I'm not farming, I care for the land and the dead. The carriage trail taking you through the ten-acre wood—a young hardwood forest heavy with black cherries and locusts—is laid with crushed stone. But during its first years of operation, the trail was not managed in any real way, just a foot-worn grassy path that meandered through the forest and on to the burial sites.

Then came our first late-winter burial. When the family called to schedule the funeral, the ground was in a state of rapid softening after a short stint of snow the week before, then rain, and then several days of unseasonable temperatures in the sixties with no wind or sun to dry things out. It would be hard to imagine a more perfect recipe for mud.

Cars are permitted in the burial ground only for those who cannot make it on foot. Luckily, no one in the group needed transport, and the trail would be spared. Still, to get to the burial site from the parking area required, for most people, navigating the kind of molten ground that could cause you to lose your footing or possibly a shoe or, even for the more agile, fighting a sinking force with each step down and every spring forward. Before the family arrived, I silently wept in my car (did I mention my predilection for worry?), anxious about how this would all play out. Predictably, everyone's footwear was filthy and damp, but in the end, no one appeared particularly distressed by the inconvenience and mess of it all. In fact, while the mud had reconfigured the choreography of the procession, and it was awkward to observe, it had also gathered everyone's attention, focusing our awareness squarely on the path and onto where we were headed.

At its most basic, green burial is about rendering the dead to the earth in the simplest way possible. In practical terms, bodies are buried in biodegradable containers at about three and a half feet deep, where a flurry of active microbes gather to promote swift decomposition. At its best, green burial fosters relationship between land and human, living and dead, through this straightforward and unadorned process. But to generate that kind of connection for the living in a time of gaping absence of relationship requires more than standing passively at the graveside as a body is buried in this way. In his formative *A Sand County Almanac*, Aldo Leopold urged us to embrace an ethical framework that positions humans as part of the biotic community, but with every bird sketch, plant tale, or mountain metaphor, Leopold also offered us a way to see how in need we are of engaging in practices that foster our reunion with the land.[1] Participating in lowering the body into the grave with straps rather than a pneumatic lowering device operated by the cemetery, filling some or all the grave back in by hand rather than by backhoe, and journeying with the body from the car to the grave rather than arriving at the burial with the body already on-site invite us to remember that we, too, are part of nature.

I've witnessed all manner of cultural and religious manifestations of green burial that engage these practices, accompanied by collective chanting or prayer, music by way of chorus or iPhone, and human silence. Regardless of kind, every one of these burials was shaped by elemental dynamics often ignored or dismissed—that is, until they get in our way or foil our plans—and over which we often have little control. That March, it was a muddy path, but it could have easily been the squeak of the wind-lashed crown of the tulip poplars midwinter and the voluble flute-song of a wood thrush holding on, the *tink-tink* of tender drizzle on sugar maples and the crunch of their leaves underfoot, or the sunlight reflected off the back of a shovel laden with sandy ground gripped by a mourner's hand. Each of them evidence of how the elements carry all of us along.

And yet earth, fire, water, and air were long ago abandoned by science and mainstream philosophy, rendered irrelevant to understanding the nature of matter, reduced to weather or environment. Early Greek thinkers, of course, respected the necessity in their naming. Before they were categorized as elements, first by Plato who saw them as "the alphabet of nature" and later Aristotle as "basic part[s] of a whole," Empedocles called them *panton rhizomata*, the actual *roots of all* things. The farmer in me is partial to this telling, as rhizomes (such as turmeric and ginger, both of which I also farm) are horizontal underground stems—gnarled, burly footings for their tropical flair above.

It may seem counterintuitive that mud has been such a persistent conductor for this ancient, almost forgotten and unnoticed knowledge—that the elements are the roots of all things. This has been especially true for me over the past decade as I've been engaged at the farm and the burial ground—two spaces geographically less than a mile apart, situated on the same foundations of soil and substrata. While mud can be used to build pots, pipes, firepits, and floors, even entire homes, if not fired by flame or cured by the sun, mud is not physically sturdy. Mud can be chaos. Indeed, the two key elements of mud—water and air—make it slippery and unstable. Mud can take you to your knees and can also carry things—buildings, people, entire communities—in the form of mudslides. Perhaps it's precisely because mud has movement and flow that it can leave us feeling not just out of step but at our most vulnerable.

Still, mud's inner current—precarious, volatile, even perilous—has the capacity to slow things down, as it did at the burial, altering pace and rhythm, allowing us to observe in ways that we might otherwise rush through. Mud can draw the present closer to us, without the obsessive desire to reign it in and hold it hostage to our human needs. Mud can act as the kind of elemental teacher that, much like the spring ephemerals, models for us the moving, and thus, fleeting, nature of the world. Some spring ephemerals, like bloodroot, abide by the word's etymology and last for only a day. In the burial ground, the Virginia bluebells last a tad longer, briefly unrolling their cerulean carpet just before the full canopy of the forest leafs out. And when the bluebells flower, I routinely call on every tool I can to try to extend their stay—camera, video, language, both written and spoken—all futile bids to chase and capture the moment.

The Danish philosopher Søren Kierkegaard knew this chase, referring to the moment as "the merely vanishing." We may feel the moment to be dire or exquisite, but it's not an "atom of time." The human understanding of what a moment is, as Kierkegaard argued, is predicated upon our first attempt to try and halt it.[2] But *moment* is derived from *momentum,* which comes from the Latin *movere,* to move, which means that to invoke the moment is to know that it is always in motion. As Plato said long before Kierkegaard's speculations, time is only "the moving image of eternity."[3] And it does not stop for us.

At about the same time I first noticed the garlic struggling, I walked the burial ground with a man whose wife had just died. A mellow April rain had fallen the night before, deepening the ocher bark of the black cherries, darkening the way. As he traced the land over and again, the woods radiated out that pleasant and often uplifting smell emitted only after rain has fallen on dry earth—a scent so exceptional it was given the name *petrichor,* stone (*petra*) mixed with the blood of the gods (*ichor*). The smell of mud can be many things—salty, when rising from a marsh; sweet, when wafting from a bog; or pungent, when carrying the tang of leaves or algae breaking down. It can also be foul like feces and thus evocative and threatening, the furtive and fertile place out of which all things emerge, seemingly disappear, and also return. But at the heart of all mud is *petrichor*, the key component of which is an organic compound that hangs around moist soil, one the Greeks called *geosmin* (*geo* or "earth," and *osme*, "smell").

In Greek mythology, this blood (ichor) is most often referred to as golden in color, but Plato claimed it to be made of "all sorts of bile and phlegm," denoting a color surely as much green as yellow. Ichor was also imagined to be a fiery substance that, while toxic to humans, could also cause "a magical herb to sprout when it touched the ground." And while mud is obviously made up of earth, water, and air, petrichor tells us that there's fire in mud, too. Not the destructive kind ever present today, but the life-producing and generative sort, the kind that clarifies and turns that old idiom "clear as mud" on its head—the very same green fire Aldo Leopold is known for having seen in the eyes of the dying mother wolf whose life he took.[4]

When my fellow farmer proclaimed, "When it snows on mud, it's over," she didn't mean it as a provocation. She simply understood mud's basic ontology—mysterious and intelligent but also inexorable, which may explain why it's so hard to thwart or redirect it. Those last four beds I'd planted, while they'd emerged and continued to grow, were radically different in size from the rest of the garlic I'd sown before the snow fell. They didn't look diseased or even distressed—just unusually small. By the time the garlic scapes showed up in their predictably showy way, shooting upward only to grow back down toward the soil and then take a corkscrew detour around themselves to form the most elegant of curlicues, I knew those four beds would not produce much of a harvest. It was mid-June by then, and the garlic had about a month or so to redirect its energy away from the leaves back down to the bulbs. With such meager growth and tiny scapes, it didn't seem likely those bulbs would ever size up. That much I was right about.

Maybe it was all the water. Maybe it was the cold temperatures from all the moisture and ice that interfered with the robust rooting that needed to happen in the fall to give the garlic a fighting chance in the spring. Or maybe it's just something about the vitality of mud when blanketed by a chilled, constricting energy while daylight is dwindling that signaled the garlic to also slow its flow. If that were the case, it would be impossible for me to expect a different outcome, because mud has a timetable all its own.

In this way, mud has a lot in common with grief, which can be found in ample supply in both the burial ground and the farm field these days, both moving with the same verve of wobbly bends and curves and with the same propensity to draw things out that it can sometimes feel as if time has ceased altogether. The Canada geese's failed migration from Mill Pond—a body of water functionally separating the burial ground from my farm—is only one part of the wreckage. Families journey to the open grave while crops are compromised from dips in pollinator populations. White pine boughs are laid on top of shrouded bodies as they're lowered down while harvests are pushed back due to rising temperatures. Shovels are handed off to backfill a grave while wildfire smoke interrupts an afternoon of planting. Invasive barberry nicks the ankles of mourners as they make their way back to their cars while an arrowhead surfaces after a torrent of rain. I feel incalculable loss. Of course, to love the more-than-human world, we must also learn to grieve it, messily and unremittingly. For whatever time it takes.

notes

1. Aldo Leopold, *A Sand County Almanac and Sketches Here and There* (New York: Oxford University Press, 1949).
2. Søren Kierkegaard, *The Concept of Anxiety: A Simple Psychologically Orienting Deliberation of the Dogmatic Issue of Hereditary Sin,* ed. and trans. Reidar Thomte, with Albert B. Anderson (Princeton, NJ: Princeton University Press, 1980), 86–89.
3. Plato, *Timaeus,* ed. Thomas Kjeller Johansen, trans. Desmond Lee (London: Penguin, 2008), 37d.
4. Leopold, *Sand County Almanac,* 130.

The Human Element

CMarie Fuhrman

The young Dog is asleep on the couch. The old one
lumbers about the living room looking for a comfortable place
to rest. This house wakes up. This house goes back to sleep. Outside
a Cowbird whistles, the breeze moves nothing. The trees
    unclench green
fists, palms open to accept first light. Now the crowns of the Fir
    turn gold. Now
the first light of day touches the wild hyacinth on the slope
where the stump caught fire one spring
when we were burning needles and dropped limbs and stayed
burning for days as fire turned root to ash in its generous way. Still
the stump stands and daylight will fall
where we placed a disk to hold water for the bird people.

This is a moment of opportunity.
Spyria are making plans for pink
blossoms. Spider vibrates in her web and shakes loose
the dew. Birdsong pauses, as if waiting for a cue. And it's not
as if Alder has a choice today. Not like the Strawberry
will withhold its fruit or the white passion of its blossom. Trillium
though their petals, purpled by time, fell and decayed weeks ago, will
grow, will move toward light, seek water, host whoever
lands, offer themselves to the nibbles of Deer who stare
at me now as if to intimate that I, too, must take part in this scene.

Perhaps I am wrong.
Maybe Trillium and Alder and Birdsong and Deer don't care
if I am here. Would, in fact, thrive if I were not.
If my species were not. I suppose I could try to live
like that. Unfurl, consume, shake the dew of my own web, then go
on with the things I've been told are important to living
in this state, this country, this body. Ignore the Arnica
that only yesterday opened along the driveway. Assume
they would have bloomed anyway. Stop apologizing
to Robin whose morning reverie I interrupt. Maybe the loss
would be neither theirs nor mine if only occasionally
I swept a hand through a Bluebird sky
to show another what I'm told is mine and then not
swallow hard when they say, oh it is so beautiful. And, oh
you are so lucky. As if I had something to do
with all this wonder. This beauty. As if,
I had nothing to do with it at all.

And then I would say yes, yes and we would go back
inside to pot roast and cribbage.
I could do that. Maybe
Spyria wouldn't even notice me gone. The Thrush would still
sing a dawn song. Forget-me-nots would have no one
to forget. But then, when I bend to each of these
beings, when I pour water to the root of Clematis
that has struggled for years to make its July climb up the
        blackened stump,
when I trail my hand over tops of Skunk Cabbage as I walk to the grave
we made beneath the Mountain Ash, I think of all I would lose
if I stopped believing
that somewhere in this same life I share
with all beings there exists a mutual recognition
and the Skunk Cabbage felt my touch, the Hyacinth heard my
        morning greeting.

Yet,

I cannot risk being wrong. Just as I cannot imagine ignoring my lover
when we meet in the kitchen to boil water, not tussling
the wild hair of my friends' children. Ignoring the wags
that respond with glee at hearing their names and good Dog.
The same life which responds in kind to me
in the human, lives in the Spider, her web, the dew. The wild
Strawberry. The throaty Grosbeak song. Call it respect. Reciprocity.
Recognition. Generosity toward kin that moves me. For bending
    the Mountain
Ash to see her blossom, moving the Snail from the trod path
to deeper detritus, placing my hand on the thick skins of
each Tree I pass is acknowledging life itself.
And acknowledging life of any kind, admitting its sacredness,
merely admitting its alive—
is acknowledging oneself.

(What kind of death might we suffer if
our bodies were used only for another's pleasure,
if we were mere resource, ignored when we didn't suit a
    perceived need.)
How lonely must Earth be for the press of our bodies
against her, how forsaken the Stars for lovers
to count them and for warriors to fight by them.
The songs of all our ancestors
for Rain, for Corn, for Salmon to return. The thanks given
for one life to another for sustenance. Even the simplest grace...

Wake up, I say—
standing beneath the old Pondo, Pine Squirrel in the boughs.
    Good morning, I say
to the Cowbird on the rail and these gestures alone enliven me.

Good morning to the dew petaled Strawberry, to the Thrush
only my attention touches. And could this be the reason for my kind?
The gift of attention, celebration that we bestow?—
(Could it be the human element is also necessary?)
Our Earthly purpose, the mere act of giving
love to that which we took no part in creating—

Go Light

Gavin Van Horn

Baja California Sur, Mexico

D awn meditation, perched cross-legged on a mesa, I absorb all the elements present on a vertical: the rocky earth of the cliff's edge, the expansive salt water, the cloud-shrouded sky, and the fiery Sun.

For the moment, a shadow of gray gauze prevails. But only for so long. All I must do is wait as Earth rotates. I know it's a perceptual deception: my anticipation feels as though it exerts a summoning force, as if my will were bound to others who are waiting. Waiting, waiting, waiting. A tremor of doubt passes through my head, some electric current of fire across a ganglion of synapses: Will it rise today? *Of course, it will.* But will it?

The elements preceded the planets. And in an inscrutable way, reaching back to before times, so did we. We are made of what we return to: matter born of stars, unearthed for a moment, and when, five billion years from now, the Sun stretches out, as those with understandings of vast timelines and cosmic phenomenon speculate, we—by which I mean the elemental molecules we call "we"—will once again be star cradled. Light.

In an elemental sense, earth precedes Earth. Water precedes oceans, rivers, and ice caps. Air precedes atmosphere. The cosmic fire, doula to galaxies, precedes it all, creating the conditions and the possibility of life from light. The Greek philosopher Heraclitus (ca. 490 BCE)—remembered for his reasoning about the elements and described as "an inward-turned observer of the world, inventor of the first philosophical genres, the thought-compacted aphorism, [and] prose that could contend with poetry"—honored fire foremost among the elements.[1]

If nonphilosophers have any reference point for Heraclitus, it might be his quasi-famous aphorism that one never steps into the same river twice. This emphasis on the flow of *relations* instead of the stasis of *things* upends object-oriented ways of thinking: a river is not a static entity but a being that is ever-changing. So, too, are human beings. But it was not water that Heraclitus lifted up as key to understanding the cosmos.

According to the philosopher Eva Brann, Heraclitus honored fire as "the root-element" because of the ways in which fire instantiates the process of transformation, acting as a kind of "cosmic currency" and "all-pervasive, everliving implicit medium that sometimes appears as an element on its own."[2] As one of the rare surviving fragments of his words asserts: "This world-order (*kosmos*), the same for everything, was made neither by any one of the gods nor of men, but ever was and is and will be: an everliving Fire, kindled in measures and extinguished in measures."[3] Fire is the transformative, ever-living force that makes all elemental motion—and thus all forms of life—possible.

It is not merely fire's ability to transform that is key to understanding this elemental, according to Brann. Viewed from another angle, fire represents a certain strife, a productive antagonism, holding everything together by a unity of opposites "*not* as their reconciliation... [but] of strenuous togetherness, of strained union, a relations that now degrades and then again vitalizes a world as a multiplicity."[4] The relations of fire to all else sets the world to "vitally vibrating" and thus "in its parts *and* as whole, a tautly vital, twangingly alive, strainingly static cosmos."[5]

I wait, anticipation building. Because of the angle of Earth to Sun, fine gradations of light arrive first, peeling back layers of crepuscular predawn morning, spilling out a wash of sparks, turning a steely cobalt sea into a shimmer of rose, peach, and saffron. Everything seems to gradually light up from within, bioluminescent.

Then—there—fire. A white-gold lip of warmth emerges, flint against the tinder of sky. I don't simply see the sunlight. I feel it. The sun's flame reaches out and warms my arms, my knees, my cheeks, my forehead, my eyes. I am touched, literally touched, by radiations of nuclear fusion ninety-three million miles away.

An outpouring of light transmuted to energy—the internal fires of all green things—fuels what is possible on this planet. This sunlight, this local fire, fills the air with a life-giving force older than Earth, coursing through me, the saguaro, the tarantula, the pallid bat, the mesquite bean, the long-limbed grasshopper, the Colorado rock crumbling into the arroyo, the hummingbird, the ceaseless sleepless sea. This generative fire combines and recombines in elegant interlocking interlocutors who speak in thousands of tongues—even through the many beings who are tongueless but not noiseless, even if only their mineral presence redounds into the infinite sky, "vitally vibrating," liplessly humming a tune older than Earth.

Pismo Beach, California

My finger hovers above the sand, hesitantly reaching toward—a shell? a creature? an alien? These castaways, seemingly everywhere, are strewn about the shoreline in unkempt bunches, matted rafts, and shipwrecked flotillas. Thousands of these 1.5- to 3-inch-long disc-shaped bodies disclose the undulating curvature of the tide line. My finger makes contact with a chitinous fin, firm but giving, that extends at a right angle above a clear oval stained a vivid indigo around its circumference. Having grown up in Oklahoma, I suffer from a deficiency of marine biology knowledge. What or who lies at my feet is a mystery. I know enough to be in awe.

Velella velella, marvelous co-op of life. Photo by Gavin Van Horn.

I later discover these creatures are *Velella velella,* commonly referred to as by-the-wind sailors, their fates wedded to prevailing winds. The translucent fin I touch with my finger serves as their "sail," secured and unyielding in its position yet making use of wind power to propel *Velella velella* across choppy open ocean. Most times, the strategy works in their favor. But if prevailing winds turn shoreward for long enough, their journey comes to an end, as it had for so many on this California beach.

I feel blown about at times, subject to prevailing winds, although I retain some small measure of control in adjusting my sail. More than *Velella velella.* I like to think so. Less metaphorically, humans have learned to harness the wind in various ways—from sea travel to electrical grids. Harnessing the elements for our own ends is arguably something most creatures do, but *Homo faber*—the ones with opposable thumbs and the civilizations to show for it— too often act on the misperception that we're the only actors in this earthen drama. We, the subjects, among a collection of objects; the only matter that matters. This has been the prevailing ideological wind in the West—religiously affirmed, scientifically rationalized— for centuries. Even as this wind drives us toward lonely shores.

Belying this anthropocentric bias, life has been shaping Life from its origins. Without such shaping—by which I mean exchange, fusion, collaboration, mutualism—we would not be here. Nor would oxygen (blessed be you, cyanobacteria, creator of oxygen-ated air), or soil (blessed be you, lichen, eroder of rock), or trees (blessed be you, fungal mycelia, for offering root systems until sky-reaching plants evolved their own). The Russian geochemist Vladmir Vernadsky (1863–1945) was one of the first scientists to recognize the protean, shaping power of life on a planetary scale— the ways in which living organisms *create the conditions* for still more diverse life. Vernadsky called this unifying, dynamic force "living matter," a force powered by a Sun that has altered living conditions of the planet over evolutionary time. He also gave it a name: *biosphere*.[6] This is the thin envelope, approximately twelve miles, top (atmosphere) to bottom (lithosphere), where life has elaborated upon itself. Increasingly in the past few decades, partic-ularly with fresh discoveries about genetics and microbial doings, there has emerged a corresponding appreciation for the symbiotic composition of our bodies—the way life shapes Life in every body.

Mergers and collaboration and partnerships among organ-isms of entirely different kindoms are the norm when it comes to life shaping Life.[7] The microbiologist Lynn Margulis (1938–2011), through her research and theories about endosymbiosis (partic-ularly the mergers of bacteria to form animal and plant cells), made this view of life central to her work. Although her theories were initially resisted—perhaps because of the strong political and cultural momentum that imagines humans as competing individuals—genetic research has borne them out.

But we need not own powerful microscopes to appreciate the way life reaches out to other life. If you are reading this outside, you may be sitting on a rock that is hosting what has been dubbed the "gateway organism" for appreciating life mergers that cut across kindoms: lichen.[8] A collaboration of fungi and algae, lichen offer a classic example of the way organisms can become intertwined for the mutual benefit of the whole; in this case, with the fungi providing rootage and mineral sustenance and the algae providing a Sun-powered source of energy.

By-the-wind sailors are their own kind of lichenlike collective. A hydrozoan colony, the sailors embody the paradigm of life shaping Life. Three different kinds of polyps, with specialties in defense, feeding, and reproduction—constitute *Velella velella*. These hang from a chitinous, water-repellent plate with gas-filled pockets of their own creation (their raft, if you will). The polymorphic colonial arrangement means that nutrients are distributed evenly among the polyps, which share tissues with adjoining polyps, forming an "individual" by-the-wind sailor—less a single sailor than an entire crew. Also physiologically noteworthy, some polyps host microscopic, photosynthetic unicellular algae (zooxanthellae) in their tissues. By virtue of this arrangement, the sailors have a backup, light-powered food source in case the zooplankton pickings are slim at sea.

The encounter with *Velella velella* evokes in me a double dose of astonishment.

The first is due to meeting a creature by happenstance, on an ordinary walk on the beach, with whom I had no idea I shared this planet. Naturalistically inclined, I tend to get familiar with what is around, lulled into a sense of general knowledge, and then *bam!* a creature from a dream, a being that looks sprung from the pages of a fairy tale or another planet's bestiary appears at my feet, reminding me of how little I know, how many creatures I have yet to—and may never—encounter. At my feet is mystery.

The second dose of astonishment comes from discovering how calling *Velella velella* a creature—as though that word were a singular noun—is a misnomer. By-the-wind sailors are a co-op, a collective of various forms of life joined together to create an indigo-dipped hydrozoan that navigates vast oceans without a central nervous system. If I didn't think it would cause others to worry over my sanity, I should walk around slack-jawed all the time because of collective creatures such as *Velella velella*.

I initially stumbled across by-the-wind sailors, this astonishment, without any idea that these were one of Life's many experiments. But here's the open secret: every being, every single being, is a collective, a collaboration of partnerships. The biologist Scott F. Gilbert has been outspoken about this, highlighting that the notion of an individual self is a conceptual construct not a biological reality. Humans are unexceptional in this respect, for, as he points out, 90 percent of the cells that human bodies comprise are bacterial, and metagenomic sequencing reveals that the human gut is "a persistent partnership with over 150 species of bacteria," with around a thousand major bacteria groups in our gut microbiome. Instead of individual selves, we should be regarded as "holobionts," a multicellular eukaryote plus its colonies of persistent symbionts.[9]

Such relationships between kindoms are found everywhere among species, fascinating worlds within worlds of startling chimeric composition—from coral reefs, to nematodes, to termites who can digest the cellulose in wood only because of their gut bacteria (who themselves are composed of five different species), to promiscuous fungal associations within forests, to Hawaiian bobtail squids and their luminescent microbial symbionts.[10] "For animals, as well as plants, there have never been individuals," Gilbert and his colleagues assert. With a nod to the "gateway organism" for understanding symbiosis, they conclude, "We are all lichens."[11]

Lichens and coral reefs, Australian termites and bobtail squids, *Velella velella* and *Homo sapiens*—all are expressions of life constantly shaping Life. And because all creatures on this planet directly (plants) or indirectly (all that depends on plants) gain their energy—their food—from the Sun, it might be said that life shapes Life because *light* shapes Life.[12]

Baja California Sur, Mexico

And what is light? Of course, familiarity with light might lead us to assume we know what it is, more or less, for we are bathed in it daily. Try for a moment to describe it, really describe it, and we may quickly falter.[13] Particle and wave; energy field. We have our words: *Quanta. Photon. Vibration.* Light carries energy that can displace electrons. It is granular; it oscillates; it is curved by the pull of gravitational fields. It strikes our eye as different colors. A prism can refract and separate these different bands of color, as can raindrops when the angle is just right, giving us those miracle light shows known as rainbows.

Staring in awe at such a colorful apparition, we may be inspired to ask, "What is color?" and nod our heads when a distinguished theoretical physicist such as Carlo Rovelli replies: "Put simply, it is the frequency (the speed of oscillation) of the electromagnetic wave light is. If the wave vibrates more rapidly, the light is bluer. If it vibrates a little more slowly, the light is redder. Color as we perceive it is our psychophysical reaction of the nerve signal generated by the receptors of our eyes, which distinguish electromagnetic waves of different frequencies."[14] But then we might think, "Wait," and begin shaking our befuddled heads back and forth instead of nodding them. And then, wonder upon wonder, we are reminded that *all* matter, not just light, is energetically vibrating. Even the stone we think is mute. Again, Rovelli: "If we look at a stone, it stays still. But if we could see its atoms, we would observe them to be always now here and there, in ceaseless vibration. Quantum mechanics reveals to us that the more we look at the detail of the world, the less constant it is. The world is not made up of tiny pebbles. It is a world of vibrations, a continuous fluctuation, a microscopic swarming of fleeting microevents."[15]

The generative source of these vibrations, our local plasmic storm of light, lies safely ninety-three million miles away, emitting all the necessary energy and more to keep the grand experiment of life on this planet going.

I am staring up at the stars, my back held by the sand. Silica—transmogrified to glass with enough fire—can be fashioned into the lens of a telescope with which to look further into the sky. On the other end of the size spectrum, a hand lens was recently gifted to me by a friend. The first time I turned it upon a flower blossom, my mouth fell open. The living geometry, the vividness of color, the surfaces—ripples, bounce, and bump—unseen but now clearly visible with just a bit of magnification were breathtaking. Another day, roaming along the wrack and funk of tideline, I turned the hand lens toward the sand: each grain a different hue, each crystal its own prismed reflecting pool. A beach that once seemed uniform to my naked eye became a multiverse.

At this moment, lying on a Mexican beach, I have come to bathe in the night sky. The sand serves as a cradle, and my "ceiling" is covered in clusters, milky streams, and singular blazes. If I turn my head, out on my peripheral: more stars, beyond the ones accounted for.[16] I gaze into a sky spangled with countless stars, thin pinpricks of light reaching across expanses so vast, there's no point using numbers. "Can you count the sand on the shore? Can you name the stars in the sky?" Such questions were one of God's rhetorical humbling tactics for an overwhelmed Job: "Can you bind the chains of the Pleiades? Can you loosen Orion's belt?" (38:31).

No—I can't even allow my eyes to linger on the hunter's star-studded belt for long before I move onto the next bit of firelight that shines. I receive a different message from the one intended for a starstruck Job: we are stardust. We are also bacteria, algal bloom, protist, archaea, fungus. Hybrid beings, aliens to ourselves, strangers in a strange land. Yet home. Home. Home. With all our fellow travelers. We, little balls of fire. Self-ablazement on a pair of legs. We are life shaping Life. Light shaping Life. As the aforementioned Lynn Margulis and her son Dorion Sagan put it, humans "redistribute and concentrate oxygen, hydrogen, nitrogen, carbon, sulfur, phosphorus, and other elements of Earth's crust into two-legged, upright forms that have an amazing propensity to wander across, dig into, and in countless other ways alter Earth's surface. We are walking, talking minerals."[17] Walking wildness. Light shaping Life. I feather the sand out under my fingertips, considering the minerals on either side of the porous boundary of my skin. Earthy earthmovers, we are.

Still, the sky beckons. I reach up my hand to touch the light. It's an involuntary move. Rationally, I know it brings me no closer. I am a child wanting to reach back, to grip an unseen finger, a Sistine Chapel motion unfrozen. This basic gesture, this reaching out, must be tied to human curiosity, our desire not only to know but also, if we are able, to love; to touch and be touched by. Too often, the controlling grasp, the clutching fist instead of the open palm, has completed the motion. We live in times bent by an extractive desire to possess, to make *mine*.

But what of the open palm? Turning to look out in awe, at the vast, shared, unfathomable sky of distance, of light-years, of other worlds—might this humbly turn us toward Life here and now, where we can apprehend the elemental sharing of kinship in bone, breath, blood, metabolism, metaphysics—raised from earth, but never far from it, kept by gravity's embrace from spinning off into space, sharing the firelight of this one star in this spiral arm of a galaxy among galaxies among galaxies among galaxies. I lift my eyes and feel reverence for what is at my feet, what makes my feet possible. I am Earth bound and Earth liberated. I cast my shining eyes above so that each step on the ground becomes more precious to me.

I am running fingers through cool sand, tapping across my kinship with stardust. My mineral skin, aswirl with other lives. And these distant stars seem to beam well wishes from afar, across black night. When I occasionally drift beyond the gravitational pull of my mind, I hear a cricket trilling in the dark—*pweet pweet pweet*. Let me tell you something, earthling to earthling: this sound makes all that dark around me light.

Morro Strand State Beach, California

At my feet, a pile of by-the-wind sailors. A gull plucks one from the sand, trundles to a freshwater stream, washes and moistens the hydrozoan and swallows them down.

What of this elemental life? We always are living it, being lived by the elements, these essential, powerful forces of earth, air, water, fire. The question, I think, is whether we are conscious of it; and being conscious of it, what do these forces, responsible for but indifferent to individual well-being, ask us to be conscientious of? How can we live in appropriate reciprocity, giving something of ourselves in gratitude for the forces that gift us with life? If we can't ever properly return the primal gift of the Sun's fire, then what steps, what orientation, what North Star shall we set our course by as we sail our elemental vessels into the unknown seas ahead?

"To climb these coming crests / one word to you, to / you and your children: / *stay together* / *learn the flowers* / *go light*," admonishes the Buddhist prose-poet Gary Snyder.[18] He seems to know—as we gaze about at the unraveling of living beings, standing at the foot of a mountain of wicked problems—that we're up against egoic monsters of our own creation. So he offers a few words of advice: on community (*stay together*), on collaborating with nonhuman kin (*learn the flowers*), and a bit of instruction for how to do that (*go light*). Best keep things simple and close at hand. No need to come up with salvific plans, technological conquests, or ride into space in search of new worlds when the old problems remain intact.

Go light. I once thought Snyder meant traveling without un-
necessary burdens, unencumbered, taking only what one needs
and no more. Go light. As he says elsewhere, we "must try to live
without causing unnecessary harm, not just to fellow humans but
to all beings. We must try not to be stingy, or to exploit others.
There will be enough pain in the world as it is."[19] I'm no longer cer-
tain that traveling simply was the only meaning Snyder imagined
when he placed those two words together, *go light*.

As a scatterer of darkness, as a force of life, light holds a cen-
tral place in religious stories and philosophical teachings around
the world. From divine presence to the Buddha's enlightenment
to Plato's Allegory of the Cave, the comfort and clarity of light
provides powerful metaphors for the journey from ignorance or
delusion to clear-sighted wisdom. One stunning example, in which
light is equated with consciousness itself, can be found in the
"Great Forest Teaching," a seventh-century BCE Indian scripture
from the *Upanishads*. The text recounts an exchange between the
wealthy King Janaka and the renowned sage Yājñavalkya, in which
Janaka fires a series of questions at the sage, beginning with the
question "What light does a person have?" The sage responds with
the obvious: "The sun. By the light of the sun, a person sits, goes
about, does his work, and returns." But Janaka presses him further.
When the sun sets, then what? The moon, of course. And without
either? Fire. And without fire, then what light does a person have?
"Speech," the wise sage replies, noting that even when one can't see
one's own hand, a voice can light one's way. But the king is not yet
content. In the absence of all those—sun, moon, fire, speech—what
light does a person have? "The self," the sage responds. "It is by
the light of the self that [a person] sits, goes about, does his work,
and returns." The "inner light" of consciousness allows a person to
perceive the world; it illuminates what can be known.[20]

Perhaps Snyder's directive to "go light" could be characterized
as a joyful elemental response: to join our consciousness—our
inner light—with the greater luminosity of Life as it is expressed

through our fellow earthlings. To touch and be touched. To keep awakening to the elemental wonders that sustain us. Go light.

gratitudes

I would like to extend deep thanks to the following beings of light who offered illuminating feedback as this essay came into its final form: Sara Crosby, Margo Farnsworth, Frankie Gerraty, John Hausdoerffer, Bruce Jennings, Josh Mabie, Daegan Miller, David Taylor, Heather Swan, and Missy Wick.

notes

1. Eva Brann, *The Logos of Heraclitus: The First Philosopher of the West on Its Most Interesting Term* (Philadelphia: Paul Dry Books, 2011), 4.
2. Brann, 53, 58, 60.
3. Brann reminds readers that all of the "fragments" that survive of Heraclitus's words are found in other texts, typically later Greek philosophers who quoted his words, sometimes to deride his views or interpret them as supporting their own ideas. We have no original document to check for accuracy, no collection of sayings as a single work to point to, although his aphoristic style does lend itself well to short quotations.
4. Brann, 80.
5. Brann, 89.
6. Vladimir I. Vernadsky, *The Biosphere* (New York: Copernicus, 1998).
7. I here intentionally adopt the word *kindoms*—instead of *kingdoms*—to highlight the lateral hybridization of life-forms, as relatives, as oppose to the monarchical and strongly divided hierarchies implied by the latter term.
8. The phrase "gateway organism" comes from Merlin Sheldrake, who himself has written one of the more fascinating books on fungal symbioses and their implications for thinking about life: *Entangled Life: How Fungi Make Our Worlds, Change Our Minds & Shape Our Futures* (New York: Random House, 2020). The quote is from p. 73, amid his fabulous chapter on lichens, entitled "The Intimacy of Strangers." Sheldrake writes, "It is no longer possible to conceive of any organism—humans included—as distinct from the microbial communities they share a body with.... Our bodies, like those of all other organisms, are dwelling places. Life is nested biomes all the way down" (91).
9. Scott F. Gilbert, Jan Sapp, and Alfred I. Tauber, "A Symbiotic View of Life: We Have Never Been Individuals," *Quarterly Review of Biology* 87, no. 4 (December 2012): 325–41.
10. In addition to Gilbert, Sapp, and Tauber, for further examples of symbiotic relationships and their key role in organismal development, evolution, and immunology, I would point readers toward the remarkable anthology *Arts of Living on a Damaged Planet* (Minneapolis, MN: University of Minnesota Press, 2017), edited by Anna Tsing, Heather Swanson, Elain Gan, and Nils Bubandt, and particularly to the essays by Donna Haraway, Margaret McFall-Ngai, and Scott F. Gilbert.

11. Gilbert, Sap, Tauber, "A Symbiotic View of Life," 336.

12. The astute reader might object here, citing as exceptions chemosynthetic organisms living near hydrothermal vents on the ocean floor or nonphotosynthetic plants, such as ghostly pipestems. True enough that there are other means than photosynthesis to provide fuel for living beings, but in both these cases it might be argued that the Sun provides a proximate source of energy. In the case of pipestem, they rely on fungal relationships that tie their nutrient intake to trees. For chemosynthetic organisms, one could reasonably point to the fact that, without the Sun's heat, there would be no liquid water (Earth would be a frozen ball of rock and ice) and hence no hydrothermal vents for extremophiles that rely on chemosynthesis.

13. If you feel flummoxed by the question, you're in good company. Einstein himself, late in his life, admitted, "All these fifty years of pondering have not brought me any closer to answering the question—what are light quanta?" Perhaps St. Symeon, an Eastern Orthodox monk who lived in the eleventh century, was right to cast a broad net: "O Light that none can name, for it is altogether nameless. O Light with many names, for it is at work in all things.... How do you mingle yourself with grass?" Both of these quotations can be found in Bruce Watson's excellent historical account of the human fascination with light, *Light: A Radiant History from Creation to the Quantum Age* (New York: Bloomsbury, 2016), 191, 193.

14. Carlo Rovelli, *Reality Is Not What It Seems: The Journey to Quantum Gravity* (New York: Riverhead Books, 2017), 60.

15. Rovelli, 132.

16. We can see faint stars better when not looking directly at them due to the distribution of rods and cones in the human eye. The technique for using one's peripheral view and thus relying on black-and-white sensitive rod cells is something astronomers call "averted vision."

17. Lynn Margulis and Dorion Sagan, *What Is Life?* (Berkeley: University of California Press, 1995), 49.

18. Gary Snyder, "For the Children," in *Turtle Island* (New York: New Directions, 1974).

19. Gary Snyder, *The Practice of the Wild* (New York: Counterpoint Press, 2010), 4.

20. This story is retold in Evan Thompson, *Waking, Dreaming, Being: Self and Consciousness in Neuroscience, Meditation, and Philosophy* (New York: Columbia University Press, 2015), 2–3. Thompson goes on to describe different kinds of consciousness, including dream states, which Yājñavalkya explores in the Upanishadic text.

On a Saturday in the Anthropocene

Elizabeth J. Coleman

On a Saturday in the Anthropocene,

as I walk in the light of a two-rivered
island to my post office, I mourn

the last typewriter repair shop
in New York going out of business;

mourn that this moves us further
from letters, from connection,

from writing home.
I mourn that it's so warm

monk parrots nest in Sheepshead Bay,
lovely as that sight is, mourn

what we've done to birds:
for 150 million years they saw

their reflections only in the sea.
Then I notice a fire escape

on a two-century-old building
casting a soft shadow; I see wheels

on a bicycle that, like meditation,
seem to slow time. I remember gorillas

stay up all night to groom their dead,
and reading about a woman in Ohio

who gave every building in town
a new coat of paint after being laid off.

At my post office, endangered too,
I avoid the self-service kiosks, wait in line

for a human. A clerk waves me over
with her smile, asks where I've been.

She tells me about a cruise she's taken
with her mother, describes the buffets,

the turquoise of the ship's pool.
Now I'm smiling too. *What's your name?*

I've been meaning to ask for ages.
Grace, she says, *I thought you knew.*

In the Magically Mundane

Yakuta Poonawalla

I n the midst of the concrete jungle of Pune, my beloved birth city in the western part of India, there is a terrace. When my bare feet touch the old, worn-out ceramic mosaic tiles of light pink, blue, and white, memories created here come flooding back, allowing me to sift through them contemplatively. The cracks on my heels are shaped by the dust, grime, and roughness of these tiles. On hot summer days, Ma would caution us when we would run upstairs to play. She would say, "Chatku lagi jaase!" (You will get a burn!)—as if the tiles were on fire. In the monsoon season, "Phisli jaase!" (You will slip!)—as if the tiles were a slide. Over the years, the fierce rays of sun and powerful monsoon rains have stained the waist-high concrete walls along the edges. It is at these edges that I have stood for hours on end with unbridled imagination, trying to understand, sense, and feel the connections between us humans, the natural world, and the elements around and within us. I call this my liminal space—I have learned here. I am learning here. I will learn here.

Named Neelkanth Hill, our apartment building has stood strong beneath this terrace since the 1980s and is one of many in the housing complex, where human and more-than-human beings find their routines and rhythms entangled and inseparable. Thirty-six years ago, Ma and Daddy bought their first home on the sixth floor, just below the terrace, with great excitement—the construction brochures promised a view of Parvati Hill from the terrace and a garden with a fountain, allowing residents to feel closer to what Ma calls *jannat* (heaven) on Earth. Alas, the surrounding urban development quickly blocked Parvati from view, with buildings so close to one another that you could look into your neighbor's home and know all the ingredients they used to cook dinner, or watch a TV show with them, without leaving the comfort of your own home. Over the years, I found the daily shows on my terrace to be far more illuminating than the dramas unfolding through my neighbor's window, turning the so-called mundane urban city life into magic.

I am in Pune on a short break from my work in an urban national park in San Francisco, California, where I create nature-based programs to help local communities experience oneness, love, healing, and what Ma calls *noor* (divine light) in nature—supported of course by miraculous redwood trees, coastal dune habitats, and the Pacific Ocean. I am back in India with a need for deep rest and come with a longing for familiarity but also new observations about life. While at home, I am drawn to the terrace like a magnet, even at ungodly hours.

Even though my small, rather dull rooftop is bare of breathtaking nature like the parks of California, it links me to something bigger, a higher power that wraps me in soft, warm, and comforting awe—the reliable rising and setting of the sun and moon, every day. The smoky, featureless, yet expansive sky where stars and constellations play hide-and-seek. The snatches of fresh air during brief lulls in the traffic. The comings and goings of urban birds and other creatures, their languages forming an integral part of the aural landscape. The plant or two that somehow grow through cracks in the walls, fed by the abundant sunlight and porous cement. And the precisely planned activities of daily human enterprise, undeterred, undisturbed, and I daresay unaware of this parallel universe.

As a child, my entire world thrived within this tiny radius made up of the premises around the buildings in our complex and the terrace—both serving as social, religious, and multicultural hubs, where families organized birthdays, weddings, funerals, and other cultural celebrations, often adding a colorful *shamiana* (an outdoor awning to protect from rain and sun). When life's dilemmas troubled Ma, her response was to point upward (toward the terrace) and say, "Upparwalla che." (There is God up above). Perhaps that was the reason my neighbors would make the pilgrimage up here, spending quiet time in reflection as if it were their church, temple, or mosque. During summers, it often looked like a food market when neighbors would lay out *papads* (flat lentil crackers) to dry in the sun, jars of pickles for fermentation, or large steel plates full of grains for winnowing. I wondered how many of the residents paid attention to the fact that it wasn't just us relying on the sun, air, water, and open space here and that the terrace was a center for harmonious gatherings of many other species too.

I am reminded of the many holidays when my terrace turned into an art studio, and I'd source inspiration for my drawings from the only grove of large trees I could see in the distance, home to species such as Peepal, Gulmohar, and Neem. Blown away by my science teacher's lecture about human reliance on trees and their destruction in my city alone, I fell head over heels in love with trees and started to make a fantastical game plan to transport all those trees to my terrace to protect them from further harm and to start my own nature school here. My empty drawing sheets were later filled with stick figures of all shapes, sizes, and colors, holding trees in their hands, and other scratchy designs that were meant to look like birds, fluffy clouds, soil, and water. Over the years, dish antennas, transmission lines, torn paper kites, and tall concrete buildings have changed this view, but the trees remain, joined by parrots, pigeons, crows, and sparrows who now straddle the electric cables at dawn and dusk everyday—their presence a constant reminder that this space has always been a hub for nature too, pulsating with life of different kinds, dissolving any boundaries and borders created by us humans.

There were many other first loves on the terrace. I fell in love with the colonies of honeybees who found their way here to build their honeycombs. Even though their phenomenal architectural design was mostly limited to the terrace, many neighbors found them to be pests and would request a group of local men, considered experts, to come in with handmade tools and neem leaves to smoke the bees out and away from the comb. Afterward, the honey was sold to the households who requested the service. Another time, a colony of bats built their cozy roost in the same location. On important prayer nights, Ma would wake up at midnight and use the carpeted floor of our living room for *namaaz* (prayer). One night, Ma's shriek woke me up. A curious (more likely lost) bat from this roost had entered through the open window and landed on her back! She flung the *dupatta* (scarf) that covered her head in an attempt to shoo the bat away. I'm sure this nocturnal neighbor had no intention of troubling Ma, who was deep in prayer, but I couldn't help but chuckle at this interspecies interaction. My young heart was drawn to darkness through such encounters, and it is still astounding to me that when I sleep, a whole other world is awakened.

Every morning, the sun finds a way to rise through the cluster of tall buildings where a black kite, a medium-sized raptor, flies in from the distant Jacaranda tree, places herself firmly on the metal fence, and investigates the area for food. I gather courage during my morning jaunts and go closer to her. As the light brightens, there is more clarity—for her and for me. She allows me to come so close that I can see myself in her. How she grips the edge of the fence with her talons, just like I grip my wooden gymnastic rings on the other side of the terrace. I can see in her a focused frown, like the one I have on my forehead. She quickly spreads her wings in attentive motion when there are signs of food and just as quickly folds them away when the catch of the day requires more patience. I observe and count her black and brown feathers and the patterns they form. If only I could attach myself to her and fly, I would have a new understanding of the shared air we breathe and the life she leads. I am so close I can sense her breath. In the wide bounty of her presence, we are connected and become one. A riot of bird songs in the nearby Peepal tree starts to call for my attention. So does the *azaan* (Muslim call for prayer). Packs of street dogs take the cue from the *azaan* and start howling. The crows and pigeons outnumber the smaller birds with their cawing and cooing. When the big orange ball of sun is fully visible, all activity hits a crescendo. Vehicular sounds join in. Pressure cookers in kitchens below start to rattle and whistle. If I were the conductor of this orchestra, I would have asked the vehicles to turn it down, and the crows to contain their excitement a bit, so I could enjoy this magical fusion performed by all the various morning routines.

When the kite leaves, I scan through the fence to see what she sees: building windows decorated with clotheslines and sills that store everything from pickle jars to plants, toys, and other paraphernalia; a ground full of rubble; and the barren land that was meant to be the fountain garden that Ma and Dad were promised in the building plans—where fresh soil, plant, and bird life could have brought peace and wellness for the families residing here. Sadly, the structure for the fountain lies dry, silently testifying that a *jannat* (garden—and maybe, in this case, a small piece of paradise) may just have been possible here if it weren't for land disputes, and lack of funds, time, and interest. Dad has spent many years serving on the Management Committee of the housing complex, and despite many hopeful attempts, his plan to revitalize this space has been rejected. "It's not a priority," they say. He continues to hold onto this dream and says, "Insha'Allah, one day" (If God wills it, one day) when I inquire. For now, he, like many others, exercises on the terrace, where he greets the kite, the rising sun, and all the other characters of the orchestra, who unbeknownst to him have become his regular companions. In the places where bees, bats, and kites have tried to create homes but have been thwarted by us, a middle-aged neighbor, head always covered with his white and gold *topi* (religious headgear), is caring for his potted plant. From the corner of my eye, while seemingly busy with my movement practice, I watch him conversing with his plant, gently observing the leaves and cleaning the dust that has settled on them.

One afternoon, I found S sweeping the hot terrace floor, her little son hanging on my gymnastic rings with a mischievous smile. S served our homes by going from door to door to collect our garbage and clean the common areas. She came from the Kashewadi slum south of the terrace—home to underappreciated yet essential service workers who many residents in my building depended on but who themselves lack access to basic necessities, schooling, or playgrounds. S's salary wasn't enough to send her son to school, so he accompanied her. His pastimes included playing with the elevator buttons, supporting S with the garbage collection, and running up and down the stairs. The terrace was his treat. I recall S telling me, "Idhar koi chillanewala nahi hain. Yahan aaram se khel sakta hain, duniya, aur bade khwaab dekh sakta hain" (There's no one to yell at him here. He can play freely, see the world, and dream big). I wanted to ask what his dream was, but I held back. Our society had stolen many of their dreams already by giving them names such as *kachrewalli* (of trash) and *kachrewalli ka beta* (son of trash) that would exclude them from many aspects of life. I noticed how S's son had a heightened sense of wonder. Hanging upside down on the gymnastic rings, he pointed in directions where the birds flew and followed lines of black and red ants on their various missions along the cracks. In many ways, he already had the *buddhi* (wisdom) of seeing the joy in the little things—an elemental way of being that is threatened in our overly scheduled and excessively materialistic lives.

Thousands of miles away, many of the San Francisco–area communities I design programs for in the park have had their dreams stolen too—having lost or been forced to leave their homelands elsewhere due to wars and poverty, seeking respite, refuge, and belonging in America, where the people, food, air, water, and soil feel alienating. Racial, social, and environmental injustices have pushed many others to feel excluded and isolated even though they were born and raised here. The park has become my new terrace, and I hope my programs are providing such a space for all. Here I guide communities in cultivating relationships with regenerative cultural and natural elements around them. A wide range of simple habitat restoration activities, culturally relevant interpretive and environmental education programs, and recreation and wellness programs offer a window into seeing nature as part of us. We've transformed park trailheads, parking lots, overlooks, and picnic benches into spaces where our communities have celebrated Día de Muertos (Day of the Dead) by creating *ofrendas* (altars), lit *diyas* (clay lamps) as a symbol of hope during Diwali (festival of lights), drawn henna art inspired by the flora and fauna of the park during Eid (celebration at the end of the month of Ramadan), and sang hymns and devotional songs with interfaith groups, all on a stage provided by the thriving coastal habitat.

I am reminded of an activity that was inspired by the symbolic meaning of Tibetan prayer flags—introduced to me during time spent in the Himalayas. A bowl with small pieces of paper with the words *air, water, fire,* and *earth* written on them in English (or another language spoken by the community) was passed around to participants, inviting everyone to pick an element and embody it for the rest of the program. A Yemeni grandmother picked the element air at a program on Alcatraz Island one day. When asked why, she said that her large family lived in a tiny studio apartment where she had learned all their breathing patterns. I wondered if her family's breath reminded her of sweet memories of Yemen. I wondered if she could tell the state of their minds and hearts by just knowing their breath. She said that the air in her neighborhood didn't feel peaceful. I wondered if she could sense the fear in her grandchildren's breathing just before they left for school, knowing that they would have to walk through the drug, trauma, and poverty-filled streets of their neighborhood.

After enjoying a walk around the island, we settled in the rose garden, where picnic blankets were casually converted into *masallahs* (prayer rugs), turning the island into a temporary mosque for the women. Through prayer and devotion, they were coming in close contact with fresh soil, plants, and bugs on the ground. The bay waters and the nesting seagulls watched with full attention as this powerful human-nature interconnectedness was taking shape. On the boat ride back to the city, the grandmother and I sat next to each other, watching the water create ripples. As we neared the city, she said that the gardens of Alcatraz reminded her of her own garden in Yemen, of rose petals they used for cooking, and how the Arabian Sea was as beautiful as the San Francisco Bay. Becoming aware of this elemental commonality, the air felt peaceful to her. "I am now this air and water," she exclaimed.

Over the years, I have had many participants express how parks have become places of affection, longing, and returning for them when they are able to smell, see, sense, or hear something that connects them to their own memories. Planting a plant or dispersing seeds awakens the memory of someone or something; the songs and migration patterns of birds remind them of their own journeys and ancestral stories. As unsafe, polluted, and cramped spaces on land become more common—and loneliness, estrangement, and climate migration rise worldwide—spaces like my terrace, and parks and nature-based programs across the world, can be profound and positive tools to shift the ways we build and rebuild our communities. My role in the parks is merely that of a facilitator—to show that the elements of nature are available to us everywhere, all the time, and that attending to them provides a way not only to recognize the magic they bring but also to understand ways to nurture them, increase their presence, and in return soothe the urban angst we all feel. I wish for my communities to smell flowers, not fumes. For them to feel joy, not fear. *Insha'Allah!*

Back in Pune, it is my last hour on the terrace before I leave for the airport. I'm at the edge, in my liminal space one last time, and so are my emotions. The colors of dusk and all my human and natural friends of the past few weeks are back and busy with their evening activities. On the terrace across from us, a friendship grows stronger as the girl I have observed daily chats away joyfully with someone on her phone. A young man sits with his dog and mindlessly fiddles with his phone while waiting for the sun to set. Two young boys fly paper kites. A neighbor is sitting cross-legged and winnowing wheat. I join her and love how my hands trail through the mounds of grain. The evening orchestra begins. The crows caw. The parrots whistle. Just then, a mother and her four children come running to the edge to watch the sky. They're looking for the moon to signal the beginning of the month of Ramadan. In a few brief moments, they find the crescent and the orchestra expands to include their excited voices. The *azaan* calls. The family lays out a large *masallah* and begins to pray.

Even in this last hour, my terrace feeds me. It may be bare in nutrients, but it still acts like an umbilical cord connecting me to all life and the elements around me, nourishing my muscles, bones, veins, and nerves. By stepping out onto this rooftop, I was actually stepping in. In the magically mundane, together we were seeing the *noor* in nature and experiencing deep communion, reverence, and holiness.

Adobe

Leeanna T. Torres

*New Mexico is one of the oldest inhabited places in the United
States and boasts an original, authentic architectural heritage
known simply as "adobe."*
—New Mexico's official tourist website

P apa pulls out the small metal key hiding beneath the wood
window seal, preparing to open the door to his *cocherita*
(shed). Standing behind him, I wait. Eager. As he unlocks,
then opens, the door, a scent of deep, damp earth fills us. Papa's
inside before I am, already deciding on tools and gathering *madera*
(wood), leaving me to play and explore. Outside, a blazing summer
sun nourishes Papa's alfalfa fields, but inside this *cocherita de adobe*,
our bodies take reprieve from the brutal heat of summer. I playful-
ly drag my purple Payless tennis shoes across the dirt floor, lightly
kicking up fine dust. It is so *fresco* (cool) in here, in this little adobe
shed, the ceiling low enough I can nearly reach up to touch Papa's
collection of spare ropes and chains hanging from the *vigas* (wood
beams) above. While Papa works, sunlight drifting in through two
small windows, I search for spiders among the corner webs. The
undeniable coolness of adobe keeping us both content, Papa un-
derstands that I'm happier *here* than with Mama's housework. And
he lets me stay and play in the coolness of this adobe shed, with the
summer sun outside, ablaze.

The anatomy of adobe is this: earth, water, straw (or grass),
and sun. These elements combined, each crafted with an intricate

ratio, then molded into a brick shape and dried for days turn-ing into weeks, are the building blocks of my homeland. These bricks are used to build entire structures—walls, sheds, buildings, churches, and homes. In many ways, adobe connects us with an-cient tradition, but it also maps for us the melding and creation of an elemental life.

Papa never made adobes himself, but his *cocherita* was made of adobe, and to this day I remember exploring and playing inside that space of adobe. And while my own father did not make adobes, two of my childhood friends had fathers for whom this was their living trade.

According to Cornerstones, an organization dedicated to pre-serving the architectural heritage and cultural traditions of the American Southwest: "Adobe's formal beauty is matched only by its practical functionality as an accessible and sustainable building material. In the midst of environmental change, a welcomed shift is occurring as a growing number of people rediscover what tra-ditional societies have always known." I save this quote, not quite understanding why, as though it defined for me what I had not put into words but grew up knowing in my bones.

Adobe is an ancient building technique and material, but what is *between* these elements? What thread *holds* the elements of adobe together? And what does that building block of the American Southwest reveal not only about the elements but also about *mi gente* (my people) of this Nuevomexicano landscape?

One of my childhood friends, Lisa, had a father who became a well-known adobe maker. A successful business man, he made a living making and selling adobe, but rumors ran throughout the small community about his temper, his anger toward his wife and children. Lisa never spoke aloud about trauma, but still there were stories, whisperings. And like the adobe walls her father expertly constructed, a reputation was also built. But what makes a person? What constructs an entire life? Simply stories? Reputation? What is mixed in with the truth of our living, our interactions, our relationships with others?

In the Southwest, many of the traditional buildings are constructed from adobe, mud bricks holding up the walls of our buildings and homes. Not only does adobe serve as the physical construction material of our homes—elements of earth and sky combined, then molded and stacked—adobe is also a physical metaphor for an elemental life, the constructs of ordinary people doing the best they can with what they have.

Some experts will say adobe was used in early New Mexico because the materials were easily accessible and cheap. Others would say adobe came from a deeper knowing, a tradition both physical and spiritual, passed down through time and across generations.

The towering walls of our church building in Tomé, New Mexico, are made of adobe and *terrón*. The building has stood in the Middle Río Grande valley since 1750. *Terrón* is the older style of brick common in our *valle*. The blocks are literally cut from, or cut out of, sod-grass fields, or *vegas*. Adobe came later. Although it is similar to *terrón* (also made of earth and sod grass), adobes are "formed" like bricks, and thus built or created, unlike *terrón*, which is simply cut out of (or directly from) the fields. But both are earthen bricks, and the architecture of both is as much science as it is art.

Of the six main walls of the church, the construction type varies between *terrón* and adobe. When I ask Papa about it, he draws me a map on his worktable made of metal and uses welding chalk to sketch it out. He knows the history of this place, the particulars of this landscape he has called home his entire life. His soul is part adobe, and he works the land as a farmer and rancher just as his father and father before him did. Using a simple chalk drawing, he explains to me which of our church's walls are made of *terrón*, and which ones were repaired or reconstructed with adobe. And as he draws, I notice the wear of his hands, how each finger is thick with work and sun. As he draws, I realize he knows more about the church building itself than the priest or archbishop.

To know a place, to understand it at a level so intimate, that too is an elemental life.

I am watching the dances at the Pueblo of Sandia feast day. San Antonio is the pueblo's patron saint, and a ceramic image of the *santo* resides under a pine-branch *capilla* constructed for today's feast day celebration. It is Saturday, and storm clouds are moving in from the west. Dancers move and pray in the sun. My friend Luke lays buried just to the north of where the crowd of tribal members and public gather together as witnesses for the feast day. I stand against an adobe wall. All the people watch the bare feet of the dancers on the open ground of the plaza. For a brief instant, I imagine Luke's body, six feet underground and wrapped in a blanket, lifeless and barefoot, his soul dancing with his relatives, Pueblo people praying and moving in the sun. Were he still alive, would he see me here standing against this adobe wall, searching for shade? Would he invite me to his home for chile and beans and oven bread?

Just then I recall how tall Luke was, in contrast to my petite frame, and his loud laughter and endless humor. Just then I realized how much I missed him. I lean closer into the adobe wall, his memory tying me to place tying me to the physical *present*.

Nothing will bring Luke back, and I imagine his bones becoming dirt somewhere near the adobe buildings of the plaza center, where the dancers pray more than just to Saint Anthony, more than just for rain, more than for any *one* thing but for *all* things. It's always the sum, the elements combined, mapping our story, the material of our daily lives.

The dancers pray as I lean against the wall. And I admit, in that moment, the adobe offers neither reprieve nor comfort. Instead, it stands strict and straight, the wall of a Pueblo plaza house, an unpredicted yet surprising steadiness.

Adobe, Saint Anthony, and the memory of Luke do not give me healing but offer instead an unspoken element of strength. Just then. An uncompromising steadiness.

After traveling to Mobile, Alabama, one summer, I sat admiring the luscious grass at one of the city's parks. Under the wide, generous shade of an oak tree, I learned the news about Rebecca's father, my other childhood friend whose father made a living by making and selling adobe. "He died around two this morning," revealed Papa over the phone.

Shocked and saddened, I thought of his children, especially his daughter Rebecca. A breeze kicked up, and thick magnolia trees lining the park seemed to insist something of the moment.

Mr. Otero, Rebecca's dad—the adobe maker, a man who'd made a living and a business by building, selling, and crafting adobe bricks—dead. He'd perfected a living by combining the elements, and in doing so, he built buildings, legacy, a life for his family. But now he was gone. Same age as my own father.

I called Rebecca *mi amiga,* and through the phone gave her my *pésame* (condolences). Her voice crumbled into a soft weeping and between sobs she said, "This morning we made an adobe, one last adobe for my dad."

As my friend spoke through the phone, I looked out into the park at countless trees and succulent grass, flanked by dozens of red-brick and stone buildings. Green and water and stone. So much moisture, an undeniable saturation, even in the air. And all of it so hauntingly different and in contrast to my home of the high-desert Southwest.

Sitting in a park in Alabama, I realized this place was one of water saturation. But my home, the American Southwest, was a place of water scarcity.

Through her sniffles, through the deep and sudden grief in Rebecca's voice coming through the phone, I suddenly realized there were no adobes in Alabama. There was just too much water here, too much rain. I thought too about friendship and about fathers.

"Thank you for calling," Rebecca repeated, and I wondered how many more dozens and dozens of phone calls she'd receive from family and friends. And as she hung up the phone, I could not help repeating to myself, *There are no adobes in Alabama; this land has too much water.*

I thought of how one building element is built for and works in one landscape but not another. I thought about how one man might be a good father while another man is a terrible father. And I thought about how one might be a good friend versus a terrible friend. All the choices of our lives, our daily interactions. How does it all combine to create an elemental life?

Alabama's landscape held too much rain to have walls and buildings made of the elements combined in adobe. Such walls could not stand, would not endure. But in the American Southwest, adobes were a symbol of our lives, our being, our essence.

Adobe itself not only defines my home, my authentic living. Adobe creates my home, my authentic living. Alabama had its own defining elements—water and stone—while mine were earth and sky. But as I walked the streets of Mobile, live oaks lining the sidewalks, I heard, through green and water and stone, an elemental life in this landscape, too, an antiquity and longing speaking without words.

I want to ask both Lisa and Rebecca, my friends since childhood, about the craft their fathers perfected, that brilliant art of creating the bricks of elements combined. I want to ask them if either of them ever watched their fathers pouring mud into molds, forming bricks out of earth, letting them bake dry and strong in the sun. This craft of creating building bricks by combining the elements of earth and sun. This craft of creating an elemental life. Adobe.

I think about what defines us as a people here in the American Southwest, a place we call home, the landscape of our deepest *querencia*. I was first introduced to adobe as a child—it was a building block all around me, an icon of the American Southwest, an object revealing not only the elements working together but also a map of our interconnectedness.

The obituary for Lisa's father, the adobe maker and business man, read:

> At the age of about 30 he became self-employed as an Adobe Manufacture, Adobe Mason.... He was instrumental in rebuilding many old Catholic churches in New Mexico by donating his skills and adobes. He also donated adobes to Native American Kivas.... He built the first adobe structure at our Nation's Capitol in Washington DC for the Smithsonian Institute... he was a builder, he left his mark on this world and his homes and structures will stand the test of time... our father was an imperfect man living in a tough world.... He was self-educated, generous, charismatic, said what he thought, misunderstood by some, helped others and lived life the best he was capable of. No matter how hard life got, he always kept his faith and love for God and never stopped trying.

I never asked Lisa if she loved her father, even after he died. Instead, I thought of the thousands of adobes he constructed, molded, crafted. All the adobes in the valley, all the walls still standing.

Adobe—elements of our simplest existence—combined to form the building blocks of our foundation.

And while no one in this life escapes pain or suffering, aren't those elements also part of the simplest forms of our existence? Combined with joy and friendship and beauty, can we make these earthen elements into loving homes, built by imperfect people living by, with, and from these elements—an elemental life? All of it, all of it combined until it is perfected, then dried in the sun.

Traditional adobe knowledge teaches me more than this. It teaches me the hidden element binding these things together, the unspoken and unrelenting benevolence.

The anatomy of an elemental life is this: joy, pain, beauty, and yes, even trauma and mistakes, all these elements braided and blended, mixed into a life, the spiritual composition of our *almas*, our souls.

During mass, our priest recites the prayer for turning water into wine, bread into flesh, and his voice lowers in homage of the holy. Covered by plaster and paintings and decor, the adobe walls of this church have stood since 1750. I am distracted, looking instead into a crack in the wall beside me. Behind the plaster, I imagine the adobe bricks stacked by the hands of my relatives, walls still standing through generation after generation. I sit and similarly imagine the dirt floor beneath the wood at my feet. In essence, while inside this church, we are cradled by earth—*tierra*—in this God-house. We are kept by earth, whisper our prayers within its walls. It is our place, our power, our hope.

Peering into the crack in the wall, I see it there, exposed— adobe—an epiphany disguised in mud and stick and earth and water. During this ordinary Saturday-afternoon mass, as the Eucharist of our Living God lays venerated on the altar, I'm distracted and focus instead on the crack in the wall, exposed adobe subtly staring back at me. Sacred. Brown earth, fiery sun, desert water, blended by hand with wisdom and faith. God in the cracks, between spaces, ever-present. And as the priest raises the body of Christ in accordance with the ceremony of the Holy Catholic Mass, it isn't the sacrament in his hands that reminds me to bow, but the wall at my side—adobe—and I lean in, my shoulder resting on this slightly sloped wall, hundreds of adobe bricks stacked each onto another.

Adobe—the map of our stories, the material of our daily lives, earth and water, water and earth. Elements of our simplest existence, combined to form the building blocks of our foundation. The priest continues the mass, and I continue to lean into the wall, knowing always, *adobe* as a deeper prayer.

Imagine We're Ancestors Dreaming

Sean Hill

And suppose the seas rose; what if megafires,
I mean, what if blossoms come too early, blooming
before the last frost, so starvation's looming for

the bees looking for petals to slide between.
Suppose a hundred-year drought for our loves
and bomb cyclones and hurricanes for what's

held in a palm and most all our conveniences
—our modern balms or what we take to be
necessities, things assuming roles as central

as the heart and our beloveds as we lean away
from our relations with human and nonhuman
beings, ones that if tended could blossom. Too

early blooming drives famished bears to scavenge
alleys for trash alms before they can retire to dens
under fur and fat blankets, warm. Imagine massacres

or holding folks as property, presuming a narrow
definition of worthy relations to screen others out
—the point of who is kin drawn keen. Suppose seas

of change rose once and can again, I mean, suppose
heart and imagination—hope in our time to remedy
questions or statements as dooming as *what if*

blossoms come too early, blooming? Suppose
a hundred-year drought for our loves and balms.

Reflections on Nature's Mirror

Liz Beachy Gómez

At the age of seventeen, I came across a scene on the coast of Normandy, France, that piqued my curiosity. We were returning to the mainland after visiting the Mont Saint-Michel, an ancient walled village perched just offshore on an island in the English Channel. The sound of drums caught my attention—their deep, rhythmic, ancient-sounding tones echoing across the waves. I wandered off to discover the source, feeling something mysterious begin to stir inside me.

I soon saw drummers with large wooden drums set on stands along the wet, sandy shore. Everyone was dressed in white and singing in a language that was not French. Some of the group was stepping fully clothed into the ocean, carrying white flowers that they released like offerings into the waves. The scene was captivating. I could perceive something powerful happening in the mélange of drumbeats, songs, ocean waves, and flowers. Something that I didn't yet understand. This unforgettably alluring ritual at the sea spoke to me in a visceral, unspoken language. The memory of the scene imprinted itself on my heart, etched in curiosity.

About seven years later, I found myself actively participating in a nearly identical ceremony at the ocean in Brazil. By then, I had discovered why the ritual called me so intensely years prior. The sea was no longer just an expansive blue, salty body of water, but instead she was the Great Mother, powerful in her love and abundance. I experienced how this collective mother draws us into her watery womb to cleanse our sorrows, reshape our hearts, and rebirth us back into the world. In ritual gratitude, we offer white flowers and prayerful songs to honor her.

I was intensely drawn to the ocean because my heart longed for her motherly embrace. I had spent years feeling like an orphaned soul, lost and unsure of my place in the world. I passionately traveled the world—living in West Africa, Latin America, and the Caribbean—for nearly ten years, looking for somewhere that might offer a sense of belonging. After much exploration and soul-searching, I found my place in a nature-loving community based in Brazil.

Over time, the ocean herself became my mother, and the expansive blue skies my father. It was liberating to shift from a long-standing desire to feel loved and accepted by my birth parents, to *knowing* that I was cared for by the ocean and sky—and their love was limitless. I also developed relationships with the river and stone, the fire and the forest, the mud and the mist, through the spiritual tradition of Umbanda. This Brazilian blend of spiritual wisdom, ritual, and celebration of nature draws heavily from the country's rich African and indigenous roots.

When I discovered Umbanda, the diverse community and nature-centered practice immediately felt like home. I plunged into learning Portuguese, memorizing hundreds of sacred songs, and learning as much as I could about the tradition. For the following ten years, I participated in weekly ceremonies, developed as a medium, and soon began to work with the public. Throughout this time, I traveled regularly to Brazil and eventually underwent key initiations in the temple's dedicated forest preserve outside of São Paulo.

During my second initiation, the *babalorixá* (head of the temple) revealed my soul's elemental essence, explaining that I was a daughter of the ocean (Iemanjá), synchronized with the part of the sea where fishermen find their fish and other seafood. He said that, similar to how the ocean connects the continents, my work in this lifetime was to bring elements from different cultures together in a way that could nourish the community. This revelation helped explain my deep affinity with the ocean and made my life feel purposeful. I felt seen and appreciated, and I finally began to believe that I might have a meaningful place in the world.

The nature-based spiritual practice I learned in Brazil transformed both my sense of self and my experience of the natural world. I came to know sixteen Orixás, or energies found in nature, who each carry a set of spiritual principles, lessons, and archetypes. Just as the sea represents the archetype of the mother and a promise of rebirth, the flames of the fire hold the energy of movement and act as a powerful guardian. The mud and clay teach us about generosity and compassion, and the forest brings lessons related to action and following our instincts.

There are four Orixás associated with the element of Fire (spark, flames, rainbow, stone), four with Earth (desert, forest, medicinal herbs, clay), four with Water (rain, river, ocean, mist), and four with Air (wind, time, timelessness, sky). Together the sixteen energies form a cycle of life that begins with the spark—associated with creative impulse—and evolves progressively to the sky, which represents illumination.

It is said that the Orixás are a gift from the Creator, who recognized that human beings would need their help to perceive the divine. These forces of nature can be seen, felt, touched, or even smelled, and each one exudes a different vibration and sensory experience. We interact with many of them daily, like the sweet waters that quench our thirst and wash our bodies or the fire that heats our food. Stone and wood from the forest give structure to our homes, plants adorn our plates with nourishment, and time binds our days together with a consistent, rhythmic flow.

Everything in the natural world reflects some combination of the Orixás, offering a sort of codified language with multiple layers of meaning through which we can interpret nature's wisdom. For example, the sweet waters of the rivers and waterfall represent loving tenderness, and the stone represents justice, balance, and solidity—thus, a waterfall cascading over a rocky cliff brims with meaning. It teaches the vital importance of balance and structure to support the uninterrupted free flow of love. Observing more closely, we might notice how the jagged edges of an unyielding rock become smooth over time, thanks to the constant and patient loving caress of the water. Should we seek answers to any questions of the heart, we can find them in the flow of the rivers, streams, or waterfalls.

Forgiveness is represented in the rain, and we can observe how it gradually softens the hard, dry earth, creating small streams through which emotions can once again begin to flow, returning life to the ground it touches. Clouds move across the sky, blocking out the light of the sun, just like our thoughts can sometimes shield us from seeing the light that shines from beyond.

After discovering this newfound language of nature, everywhere I looked the natural world seemed to burst with wisdom. It was as if nature spoke through every breeze that caressed my skin, leaf that brushed my arm, or ray of sunlight that burst through the clouds.

Each year our local group of practitioners was invited to spend ten days in Brazil, immersed in nature in the temple's forest preserve. There we walked barefoot, slept on the ground under the stars, bathed in the waterfall, danced to the drums, and spent each day in ceremony. We were happily sent out to clean the forest and the streams, tend the gardens, or build shelters for new initiates. We spent many hours listening to teachings from the *babalorixá*, who expanded on the spiritual principles and traditional stories related to the Orixás from a deeper and deeper perspective each year.

This special place in Brazil became our axis; its influence rippled out into other natural places closer to home. The beaches, forests, rivers, moon, and stars shone with a magic imbued by the tradition and its sacred ceremonies and teachings. The elements took on new dimensions as we participated in ritual fire ceremonies, emotional cleansing in the waterfall, energetic brushing with green leaves, or submersion in the ocean. During initiations we were anointed with milk and honey, bathed in special mixes of fresh herbs and smudged with dried herbs, then sequestered in sacred huts for quiet contemplation until it was time to reemerge, reborn into new roles in the community to great celebration.

Participating in these practices and ceremonies was profoundly transformative for me. In the loving embrace of the natural world and a warm community, my feelings of being lost, lonely, and disconnected, softened and fell away. In their place sprouted a sense of purpose and deep connection—not just to a community of humans but also to a world bursting with life, wisdom, beauty, and love.

I began to appreciate how nature's beauty is also reflected within us. Our inner landscapes mirror expressions of nature, and the outer world so often mirrors our inner world. Each of us carries our own exquisitely unique elemental composition. Some of us are predominantly Fire, while others carry more Earth, Water, or Air. You may know people who are solid as rock—in their physique and the way they live their lives. You likely know others who are graceful and seem to flow like the river or ocean. Some people are full of creativity and in constant motion like the flames of the fire. What moves you? Which expressions of nature best reflect who you are and how you approach life? Perhaps you are like the flickering fire, the starry night, the meandering stream, or the tree that takes root in a crack in the rocks and grows against all odds?

In Brazil, practitioners refer to each other as "sons" or "daughters" of the Orixás, meaning that they are strongly influenced by and resonate with that particular quality of nature. This simple statement quickly communicates who a person is and how they operate in the world, without needing to describe them in detail. Often the person will have characteristics similar to the archetype associated with their Orixá. Be it the artist, hunter, healer, matriarch, maiden, philosopher, warrior, or any other archetype, the sons or daughters are typically very much aligned with their guiding Orixá's archetype and qualities.

For example, in my family we each reflect a different elemental influence. My oldest son has a strong influence of both Fire and Air. He's always ready to jump up and do something, to help out or give his opinion. He spends much of his time analyzing, organizing, and pondering big questions about the world. He's not very present in his physical body and doesn't care much for material possessions. He has the mind of a philosopher and strategist, with a particular passion for technology and innovation. As soon as his little body was healthy, he came into the world with wide eyes, ready to go. I soon recognized him as a son of the flames and the sky.

My younger son by contrast is very Earth-oriented. He's firmly rooted in his body, naturally athletic, and very much into the physical world. He is a talented storyteller who loves adventures and dramatic tales of heroism, which he has been crafting since he learned to speak. He has a sharp sense of humor and passionately loves animals. As I was giving birth to him, I could see and smell the trees around him and knew I was welcoming a son of the forest into the world.

My husband is in every way a son of stone—the archetype of the just king also associated with lightning. He is solid and unwavering, deeply dedicated to his work and his family. He tends to see the world in black-and-white terms, always centered on what is right. As a daughter of the ocean, I appreciate his strength and constancy. This blend of the elements within our family keeps things lively and dynamic and helps us appreciate one another's strengths and attributes.

Just like the rest of the world, we live, move, and process our experiences according to our nature, in our own way. In the same way that we don't expect the forest to act like the sky, or the water to condense like a stone, celebrating our unique nature can make it easier to accept ourselves and others for who we are, without judgment.

Learning that my nature was like the ocean helped me understand my strong instinct to mother and care for those around me. It taught me to respect my sizable emotions and honor the intense creative force within me that is constantly wanting to birth new ideas and projects into the world. It also helped me come to terms with my intolerance of stagnancy and routine: the ocean is in constant motion and her movement can't be bottled up or contained; she needs the space to ebb and flow.

I believe that the elements of Fire, Earth, Water, and Air and their related expressions in nature are fundamental to understanding life. They are at the very heart of creation itself—bursting forth from the depths, layered into form. Combustive qualities of Fire produce the sun that warms our planet, echoed in the passion and drive that propels us through life. The principles of Earth and her cycles of birth, growth, exploration, death, decomposition, and rebirth reign over our physical bodies. Water brings renewal, flowing through us in emotional currents. Air fills our lungs with the breath of life, circulating in the form of our thoughts and understanding.

Each of the elements is essential to life, and each reflects a part of us. Through their multiple layers of meaning, we can learn to read the signs of nature, decipher spiritual lessons, and explore the depths of our psyche. And we can work with aspects of the elements in ritual to help realign and rebalance aspects of ourselves.

Throughout my years living overseas, it became clear that despite the corrosive impact of colonialism, there are still vibrant and strong reminders of humanity's deep connection to nature. Many songs, rhythms, and ceremonies that celebrate the essence of the elements are alive and well. They continue to bind people together through connection to nature and one another. As in times past, many communities still gather to celebrate at sacred spots in nature or on special dates of the year, with hearts full of reverence and gratitude. Elders are willing to share their knowledge and help shepherd the world back toward a place of mutual respect and reciprocity.

The many challenges we face in today's cities and modern societies are both deeply personal and shared. But I believe we can create a new way forward, a path that honors our need for meaning and belonging and offers a renewed sense of connection to the natural world.

Nature continues to whisper to us, urging us to remember and reimagine our mutual relationship. She calls us to step into the forest. To gather around the fire with our loved ones in ceremony. To wash our hearts clean in the waterfall or the ocean waves. Or climb to the mountaintop to contemplate the skies and let the wind blow away our limiting thoughts. She nudges us to find the places in nature that resonate deeply and make time for them. Despite humanity's irreverence, nature persists and prevails, continuing to offer a resplendent mirror of wisdom and guidance.

What is an elemental life? I believe it includes simple rituals in nature, conducted with clear and positive intentions, that can create profound inner shifts. When we come together in outdoor ceremony with gratitude in our hearts, something mystical happens. We connect with the elemental forces. The sun, the rains, the wind, or the waters seem to collaborate with our intentions and amplify our energy. It's as if they rejoice when we gather and invite them in.

Throughout many years of both facilitating and participating in these types of ceremonies, I have seen the clouds part to shine a ray of light on our circle at just the right time, in sync with the ceremony. And a gentle rain come to bless the group just as a poignant initiation ended. I have also watched the ocean churn up intense waves for a friend who needed a deep cleansing, and then immediately quiet down for other members of our group who didn't.

Nature benevolently embraces and supports us in our healing. The fire gladly transforms our pain while the ocean blithely supports our rebirth. The rivers, forests, and meadows love to hear our laughter. The birds, butterflies, and flowers appreciate our admiration. They are here for us as allies and mentors, joyously teaching through their very existence and their example.

When we build simple ceremonies of gratitude or transformation into our lives and communities, it becomes natural to lovingly care for the river, or forest, or grasslands. Not only can returning to these practices help us heal wounds within ourselves, but they may also ultimately help restore our planet.

There is a very simple daily Umbanda practice that I love, to be done alone in quiet meditation. It involves simply sitting with a white candle and a glass half filled with water, in which all the elements are represented (Fire by the flame, Earth by wax, and Water that shares the glass with Air). After taking some deep relaxing breaths, shift your attention to each element within you. If your Fire, Earth, Water, or Air feels out of balance, simply call on its representation to help strengthen or realign that aspect of yourself. This regular practice brings awareness to how the elements move within you and what it takes to keep them balanced. It builds an intimate relationship with the elemental forces and an appreciation for their guiding roles within us.

May we increasingly see ourselves mirrored in nature's reflection, and honor the ways in which we, too, reflect the natural world. We are woven from the same elemental threads, originating in one mysterious source. The elements are our companions and guides on a winding path that eventually leads us home to that source.

May your fire burn brightly and show you the way.
May the earth be soft and fertile beneath your feet.
May your water flow freely and quench your thirst.
May the winds lift you through the air to the place beyond time.
And may the rays of the sun warm your heart and illuminate your soul.

Supracellular: A Meditation

Sophie Strand

Down by the river, I sometimes think I can scent the moment when the tide reverses and the estuary injects its salt response into a freshwater question. The air bristles, the wake of a passing motorboat cuts crisp Vs in the water. The salt in the air acts like salt in a soup: it enhances the flavor of the honeysuckle, the locust blooms, the sun-toasted cedar planks tilted against the boat-building shop. Mahicantuck, the river is called by the Munsee Lenape people, which means "the river that runs both ways."

The ocean refluxes into the river. Is it, then, still a river? Or is it a brackish finger probing into the landscape? I squint, as if trying to mark the spot in the water where opposed currents intertwine like the snakes of the caduceus. The water surface, spangled with willow frond reflections, keeps its secret. But at my feet I spot something else: the velveteen dome of a mower's mushroom pokes up from the grass. I bend down, noticing how the diminutive parasol structure disappears into the fine-grained soil. I know that although this fellow looks lonely, he is really less of a fellow and more of a swarming festivity below ground. Mushrooms are reproductive flourishes of fungal life-forms that live in soil (or wood) as threadlike, filamentous webs called mycelium. Mycelium is composed of long tubes of hyphae. One network might have thousands of different hyphal tips all capable of forking, fusing, foraging, and creating increasingly complex connections with other fungal systems and vegetal life-forms. I lightly press my finger to the mushroom, imagining my mind slipping like a yolk from the egg of my brain, into my arm, then strained through my finger into the mushroom's damp body. Deeper still, below the fruiting body, my consciousness dives on mycelial threads into the underworld. It holds tight to roving nuclei as it travels through a chain of opening and closing doorlike pores called septa.

If I had entered into an animal cell, I might have found myself profoundly confined, abiding by the strict rules of inner and outer, organelles bolted to the floor. But here in this mycelial network, I am not confined to one node. I can wander through the whole web. My ride is possible only because fungi are biologically unusual, confounding standard ideas of the cell and the self.

In 1665, Robert Hooke was the first to observe a microorganism under a microscope. It was his observations that, over the next two hundred years, informed the concept of the cell as a fundamental unit of multicellular lifeforms. The concept of the cell as a stable and foundational unit, though it applies to most of us, is not universally true. Many plants and filamentous fungi employ a type of cellularity much closer to verb than to noun. For example, while hyphae are commonly referred to as fungal "cells," the term is misleading. A classical cell can be envisioned as a discrete bundle of protoplasm with one nucleus, neatly bounded by plasma and an extracellular wall—one bead in a necklace. But fungal hyphae behave differently. They create supracellular networks. Hyphae are separated by cross walls called septa. However, septa possess pores that open and close, creating a fluid passageway through a hyphal thread. For mycelial fungi, there is no discrete bead on the necklace. The necklace becomes a flume through which organelles and cytoplasmic material flow. A single hypha can, at one time, house multiple nuclei. And given fungi's proclivity for promiscuity, those nuclei may not even carry the same genome as the original mycelial network. Fungal webs can fuse, exchanging nuclei and promoting genetic diversity.

I reach the end of a hyphal tip, bounce on a plush organelle called the Spitzenkörper that sits at the head, and scream as it cleaves beneath me, splitting off in two directions. A nucleus collides with me from behind, then another ramifies. The pressure passes my mind through the needle of another mind. I let myself fork and divide.

Many days I have sat at the river and imagined I am water. Fire. Air. Elemental. Matter flowing through other matter. A hummingbird. A sturgeon. The wet-eyed opossum under the barberry bush. The barberry bush. The Wolbachia bacteria pinwheeling through the colorless blood of the monarch butterfly. The exercise necessarily fails. But the empathy muscle it strengthens is crucial. In an age when anthropocentrism is fueling mass extinction and ecocide, it seems vitally important to practice thinking like other beings. Or even, when we feel ambitious, trying to think alongside elementals and the deep-time oscillations of entire ecosystems.

We have behaved like ordinary cells for too long, pretending there is no movement from the inside to the outside or vice versa. We have believed, for too long, that our minds belong to us as individuals. But advances in everything from forest ecology to microbiology show us we are not siloed selves but relational networks, built metabolically by our every biome-laced breath, thinking through filamentous connectivity rather than inside one neatly bounded mind. I think of the spider who, sitting like the iris inside a lacy eye, tugs and flexes and tightens her grip on different strings, creating an interrogative experience with web and with world. Scientists have likened this behavior to the activity of a brain itself, sifting through and reacting to stimuli. Each tug is a question, each returning vibration a reply. In this way spiders can sense which parts of their web attract more flies and focus their continued silk production on those areas. They can tell immediately when prey has been caught; and studies have shown that when webs are deliberately damaged, spiders perceive the damage and locate the spot, where they hurry to make repairs. Even more strangely, the extended cognition researcher Hilton Japyassú has shown that cutting a part of the silk dramatically shifts and disorients the behavior of the spider, seemingly imitating the effects of a lobotomy. This begs the question: Where is the spider's mind? Is it inside the spider's actual brain? Is it in the spider's spinnerets or legs? Is it in the web itself?

As the cognitive philosopher Evan Thompson describes:

> Part of the problem, however, comes from thinking of the
> mind or meaning as being generated in the head. That's like
> thinking that flight is inside the wings of a bird. A bird needs
> wings to fly, but flight isn't in the wings, and the wings don't
> generate flight; they generate lift, which facilitates flight.
> Flying is an action of the whole animal in its environment.
> Analogously, you need a brain to think, but thinking isn't in
> the brain, and the brain doesn't generate it; it facilitates it.
> The brain generates many things—neurons and their synaptic
> connections, ongoing rhythmic activity patterns, the constant
> dynamic coordination of sensory and motor activity—but
> none of these should be identified with thinking, though all
> of them crucially facilitate it. Thinking is an action of the
> whole person in its environment.[1]

Thinking, then, is constituted less by an organ and more by a relational process. Life is an elemental verb, stitching us into other habits of mind, pouring our ideas into morphologies and minerals better suited to navigating complex systems.

I think my mind is not just in my body. It is in my entire web. My entire web of relations—fungal, geological, microbial, vegetal, ancestral—that weave together my specific ecosystem. Sometimes, in the morning, when I call on each of these beings in a practice I loosely name Gathering Counsel, I imagine that I am like a mycelial network below ground, opening up the septa pores in my branching hyphae. I am opening myself up to a supracellular state whereby my mind can pass through my threads of relation into the minds of woodchucks, black bears, chanterelles, and juniper trees.

Given that most ecosystems are experiencing anthropogenic disruption, this extended cognition is not always painless. A year ago, my favorite lake was "managed" by the local land conservancy. This involved cutting down more than three hundred trees, some of which were home to bald eagles I had known for years. The management program decimated the local beaver population and the diverse array of wildflowers that had clung like a Technicolor crown around its shoreline every summer. Now, every morning, when I go to summon that part of my mind, I find a blank. I think of the spider with its cut web and disoriented behavior. Part of my web has been cut. I cannot flow through my whole web of consciousness. My thinking becomes disorganized. When those trees were cut down, when those eagle nests were dislodged, where did that part of my mind go? It feels equivalent to the loss of a neurological function after brain injury. I've lost the ability to use my left hand. To distinguish faces. To recover certain words.

I gaze out at the River That Flows Both Ways and think of the countless Munsee Lenape who died right here, on these banks, slaughtered by the Dutch. I think of that frayed spot of the web. I feel their absence in my extended mind. I open up my septa to the nucleus of their void. If nuclei can flow both ways through open-door cells in a mycelial network, maybe time can also flow both ways, refluxing backward from the future like the ocean sending its vein of brine back down into the Hudson and also flowing from the distant past. Can the Munsee Lenape who died here still reach me? And then, leaping forward, I ask the guardian angel of my own self: *What next? How do we bear this?*

I summon the supracellular not because I want to go on an adventure but because I earnestly want to think better. And I have a hunch that thinking better means leaking outside of the bounded cell, the individualized flavor of consciousness. In a similar vein, the Welsh *Mabinogion*'s legendary bard Taliesin recounts:

I have been a blue salmon,
I have been a dog, a stag, a roebuck on the mountain,
A stock, a spade, an axe in the hand,

A stallion, a bull, a buck,
A grain which grew on a hill,
I was reaped, and placed in an oven;
I fell to the ground when I was being roasted
And a hen swallowed me.
For nine nights was I in her crop.
I have been dead, I have been alive.

Whenever I read this poem, I think of how all poetry, all mysticism, all great wisdom comes from a willingness to leak into other being's minds. To be a salmon. A stallion. A grain. To know that while we may superficially function as individuals, we are really part of a long-term project in supracellularity whereby all our physical matter flows and recycles and recombines. To be a better salmon, be a man for a while. To be a better man, be a river. To be a better river, let yourself be invaded by the nucleus of a distant ocean.

If a cursory study of somatics shows that we think with our entire body, then how much better could we think if we did so with our entire web of wild kin? I want to think and feel and weep and grieve with my whole multispecies, polynucleated mind. I want to let the yolk of my small desires slide into otherness. I want to nucleate a symbiotic quest for a better future. Throw open all the doors in my cells. Let my river run both ways.

notes

1. Evan Thompson, "Spring Forward or Fall Back: Changing Times for Neuroscience," *Psychology Today*, May 13, 2015, https://www.psychologytoday.com/us/blog/waking-dreaming-being/201505/spring-forward-or-fall-back.

Age of Turbulent Geometry

Matthew Olzmann

Part of the contract one makes with God when they live
in Vermont, a hundred miles from the ocean,
is we're not supposed to worry about hurricanes,
and yet, our weather is changing, making plans of its own,
pinning a little red pinwheel of despair
to the weather.com map and twirling it this way.
The town where I live rests
in what meteorologists are calling the "cone of uncertainty."
I've never heard this phrase before, but it feels so true.
Yes, I think, *I'm in a cone of uncertainty!*
Maybe I've always been in a cone of uncertainty!
Maybe I've skittered between skepticism and doubt all along!
But then I remember other times when I was trapped
in a funnel of mild hysteria or a sphere of grief.
Emotions can assume many shapes;
you can find yourself in any of them.
There were days I wandered through a cube of wonder,
uncountable hours dazzled in a hexagonal prism
of delight, though these shapes often felt contained
by the larger octahedron of existential anxiety.

On the TV, the news channel has sent a solitary man
to stand at the edge of wind-kissed town and describe
the hurricane as it takes its first swing. They make him
stand there, I assume, because they do not love him.
Audiences need to know the shape of what's out there. Why?
My torus of curiosity. My tetrahedron of guilt.
We are terrible to each other more than we are kind.
"What's it look like out there?" says the anchor
from the safety of the studio. "Bad," says the reporter
from his orthotope of dread, "It looks bad."
My pyramid of panic. My ellipsoid of fear.
How can we survive all that we feel?
They say have flashlights and bottles of water.
Stay indoors and away from windows.
My pentagonal anti-prism of antipathy. My helix
of hurt. If there's a lull, they say, don't go outside.
This could be the quiet eye of the storm,
the brief cylinder of calm that passes quickly
before the worst of this weather finally arrives.

Hope Is Elemental: Some Lessons from a Vermont Flood

John Hausdoerffer

Beside Solstice Lake, Sarah Eve and I sculpt land not into the shape of ourselves, for we are here for little more than a moment, but into the shape of Solstice Lake—a spectacular reimagining of our home upon water.

—Sean Prentiss[1]

Sometimes the smallest actions awaken hope for much larger possibilities. Between July 10 and 11, 2023, eight inches of rain flooded Vermont's Northeast Kingdom. The storm caused the worst flooding the area had seen since 1927. One month earlier, my partner Suzanne had bought a modest, Swedish-red cottage on the wooded shore of a hundred-acre lake that the local poet Sean Prentiss calls Solstice Lake. Suzanne (a design and architecture writer by profession) had dreamed of transforming a weathered, midcentury seasonal "camp" into an inviting, clean-lined, multigenerational hub of kin, including her sixteen- and eighteen-year-old kids, her best friends who lived a forty-minute drive away, my fifteen- and twelve-year-old children, her brother and family a few hours south, each of our moms, and my best friends, who live a ten-minute paddle by canoe around the nearest cove. A home for all held by the elements of the lake—waters to paddle, Green Mountain air to breathe, rich earth in which to plant blueberry bushes, campfires around which to gather.

Within a month of Suzanne buying the lakefront camp, this vision of gathering a blended bunch of family and friends met the stubborn realities of the Anthropocene. What began as a steady rain transformed into a once-in-a-generation downpour. Solstice Lake's waters rose forty-eight inches, turning the peninsula community into an island without electricity, mobility, communications, or refrigeration. The lake climbed to within five feet of the cottage. Suzanne knew she was lucky. Damn lucky. One lake neighbor's home was submerged in waist-deep water, countless small businesses in nearby Montpelier closed their doors—perhaps forever—and, tragically, a man drowned in his basement in nearby Barre. Hope was hard to find.

As the lake began receding on July 12, my preteen daughter, Sol, ventured out on her paddleboard while people mended homes and lives. Sometimes we—all of us—need the playful energy of a twelve-year-old, especially in urgent times. I followed Sol's paddleboard in a dented aluminum rowboat. Sol stood tall, paddling past a couple of loons searching the silty water for fish to feed their young chick. The lake was littered with debris.

Sol paddled to various pieces of junk, cautiously knelt down, and lifted each item onto her paddleboard. She continued until a pile of saturated waste covered the whole board, knee-high. I rowed beside her, filled my boat with debris that she collected, and brought it to shore. I returned several times so Sol could keep restacking her board with new waste.

It was persistent work. But the required patience opened my eyes to the poststorm world reemerging. Robins followed loons into the damp air to gather food for their loved ones, seeming to give courage to the warblers, goldfinches, and flycatchers who eventually followed the robins' lead. A beaver inspected her lodge. On the opposite grassy shore, a deer gingerly stepped out of the forest. Soon, the cove was mostly clear of garbage.

Even as we hauled in our final load, Sol and I knew that we could not change a climate-chaotic world with one feel-good action, especially as newcomers. In fact, a good chunk of the floating debris was from a dock the previous owner had left on Suzanne's property, so some of our purpose was to relieve guilt on a local scale, unsure of what to do with our responsibility for climate change on a global scale. We knew that our cleanup was likely more about self-care—feeling useful and moving forward in the midst of community isolation and anxiety—than it was about systemic impact.

The summer of 2023 provided blunt reminders of how the elements create unlivable conditions when humans refuse to live, adaptively, with them. The elements of air, earth, fire, and water raged in response to the collective refusal of dominant societies to see and take responsibility for the consequences of our comforts—the hottest air temperature averages on record, leading to drought conditions that dried fertile earth needed for food and habitat in the Midwest and Southeast; triggering fires in various locations, from the continental smoke of Canadian forests that turned this Vermont world gray to Maui's conflagration that claimed nearly one hundred human souls; warming ocean waters, which evaporated and dropped the torrential rains that flooded little Solstice Lake. This "climate summer" revived a humbling and enduring question: could a single action—like gathering scrap from a flooded lake—inspire greater attention to and care for the elements that make life possible and that make human life livable?

Following Sol around Solstice Lake, reassured by the flutelike call of the loon, I was reminded that care, too, is an element—both in terms of care being essential to (as in something being elemental to something's core nature) and in terms of care being an ingredient of (as in something being an element of a larger whole). Care might be our clearest, most elemental state of being. After all, care presents the elemental gifts of what we can offer to the elemental needs of another person, being, or system. Care dissolves the boundary between self and world, because caring for one another and our world connects our health to the health of everything. Care disintegrates the illusory wall between selflessness and selfishness, because caring energizes us as much as being cared for enlivens us. If the Indian philosopher Vandana Shiva is right when she says that "spirituality is understanding interconnectedness and holding it sacred," then care is the most sacred act available to us.[2] Care is the path to interconnectedness, through emotional, intellectual, and community understanding. Care is elemental to being human, to enjoying kinship with our environment, and to restoring the energy we need to care for ourselves, other humans, and more-than-humans.

First, care is a key element for being human. Care can even be seen as among the evolutionary elements of being human. In his *The Descent of Man*, Charles Darwin says that "any animal whatever, endowed with well-marked social instincts, would inevitably acquire a moral sense or conscience as soon as its intellectual powers had become as well developed as in man."[3] Here Darwin goes beyond a survival view of the social instinct to suggest that "a moral sense or conscience" is an evolutionarily prerequisite—a foundational element—to becoming human. More recently, the moral theorist Nel Noddings described how care is a vital human quality, stating: "We become individuals largely through the relations to which we belong... care ethics posits relation as ontologically basic and the caring relation as morally fundamental."[4] "Ontologically basic" means that care shapes identity; it is elemental even to our individualism.

Second, care is a core element for being a member of the ecological community, to living in kinship with the elements. In Aldo Leopold's short essay "On a Monument to the Pigeon," Leopold contemplates the paradox of the human species; we can at once drive a species to extinction and mourn that loss: "To love what was is a new thing under the sun, unknown to most people and to all pigeons. To see America as history, to conceive of destiny as a becoming, to smell a hickory tree through the still lapse of ages— all these things are possible for us, and to achieve them takes only the free sky, and the will to ply our wings. In these things... lies objective evidence of our superiority over the beasts."[5] Leopold is being ironic in reclaiming the notion of human superiority. Where many before him measured human superiority by technological prowess or by anthropocentric, rational dominance over other species, Leopold says that our superiority lies in our capacity to care. He finds it quite hopeful that "to love what was is a new thing under the sun," imagining a revolution in care as our new dominant element in the human whole, justifying a new superiority.

In the same essay, Leopold cites Darwin: "It is a century now since Darwin gave us the first glimpse of the origin of species. We know now what was unknown to all the preceding caravan of generations: that men are only fellow-voyagers with other creatures in the odyssey of evolution. This new knowledge should have given us, by this time, a sense of kinship with fellow-creatures; a wish to live and let live."[6] That "wish" is the wish of care, and it is the central element in allowing a return to the humble notion of living as "fellow-voyagers," of living with a "sense of kinship with fellow creatures."

Finally, care is a vital element for restoring the energy we need to care for ourselves, other humans, and more-than-human forms of life. In a 2023 study in the British Ecological Society's journal *People and Nature*, the author Michael J. O. Pocock and team ran a randomized study of five hundred participants that measured psychometric outcomes resulting from nature connectedness and "pro-nature conservation behaviors" such as community science (formerly known as "citizen science"). They found that community science, when designed in a nature-connection environment, increased feelings of calm and joy in participants. They concluded that "nature-based citizen science is more than just a way to gather environmental data: it benefits well-being and nature connectedness of participants, and (when in combination with noticing nature activities) pro-nature conservation behaviours."[7]

So, care provides an elemental ingredient for one's identity, one's humanity, one's kinship with the more-than-human world, and one's resilience in caring. But how does a simple act of care— like Sol on Solstice Lake—change anything in the face of such massive problems such as climate disruption? Marian Barnes, in her book *Care in Everyday Life,* makes a key distinction: "We need not only to care about but also to care for."[8] Of course, I need to care about the climate chaos driving the 2023 Vermont flood. I need to care about structural political, economic, and social changes necessary to dramatically reduce the dominant world's carbon footprint—50 percent by 2030 and net zero by 2050, if not sooner. But Barnes's study reminds us that caring for the places disrupted by climate change is just as important. Caring for this little cove with Sol, in even the smallest of ways, awakens our deepest potential for human agency, connects us with our more-than-human-kin, and recharges us for the larger battles of caring about our world on a planetary scale. Often, caring about climate change as a complex, wicked, global problem—without something simple, bite-sized, and local to care for—leaves me in despair. But merging caring about (educating myself and my students about the devastating complexity of climate change; participating in larger social movements for structural change) with caring for (making sure my child is emotionally and physically safe in a climate disaster; helping her find a way to make some positive change in the face of a tragedy) leaves me feeling calm, clear hope.

As Sol and I returned from our last haul, as we finished caring for the lake for the day, I rowed across the glassy, still-muddy water. The water's restored calm buoyed the gliding Sol, paddling a board free of waste. The scene stirred in me the hope that a broader calm—emotional, social, economic, ecological, and climatic calm—would follow. Her small, elemental act in the face of the planetary disruption of the elements inspired hope in me that a larger story could emerge from something so simple. A story of an elemental life, of a life lived in kinship with, in service of, and guided by the elements.

That story expanded as we learned about this lakeside community. After returning to land, Sol and I found Suzanne getting to know her new neighbors at a community recovery party, discovering that care for others had preceded and expanded far beyond me and Sol.

The Solstice Lake poet Sean Prentiss, his wife Sarah, and their six-year-old daughter, Winter, hosted the party and offered power from their generator to all. The theme of the party reminded me of Sol's playfulness in paddling out into the lake; it was a Bring Your about-to-Spoil Food Feast. At the gathering, everyone shared something to get us through this hundred-year flood—one neighbor had plenty of water to share, others had extra food, several had trucks for pulling people off of our temporary island in an emergency, another had decades of stories about previous (though smaller) floods, making us feel less isolated in time and space. The next day, one neighbor, who years earlier had threatened another neighbor, hugged his former enemy like a brother after working together for hours to dislodge a truck from a small landslide. These acts of care reminded me of the author Lauret Savoy's observation that "we may find that home lies in *re-membering*—in piecing together the fragments left—and in reconciling what it means to inhabit terrains of memory, and to be one."[9] Witnessing this lakeside community re-member its homes, habitats, and relationships back into a whole, becoming one even beyond the "terrains" of agonistic memories, showed me that care is the glue needed for piecing together the fragments.

Care is elemental—on Sol's individual scale and on Solstice Lake's community scale. Care is elemental because each of the actions I witnessed that week were and are simply a microcosm—an element—of the connectivity of all people, all beings, all elements. And that connectivity must underlie any effort to scale care beyond a lake community, across human and more-than-human communities around the planet.

One example of the elements sustaining human community is a fifty-foot waterfall that flows into Solstice Lake, from the low shoulder of Solstice Mountain. The lake and the mountain are deeply interlinked. As Prentiss wrote: "Solstice Lake [is] our heart-home. Solstice Mountain will be our shoulders."[10] If so, then Solstice Falls are the lifegiving veins of this heart-home. The waterfall descending from the mountain replenishes the lake. Solstice

Falls offered a place for members of the community to step away from the emotional and physical labor of flood rehabilitation. I met one neighbor there who remembered when her husband proposed to her at the falls. She was the second person I had met that week who had been proposed to there. The elements of Solstice Falls care for the people of Solstice Lake.

I am a person who is drawn to unheralded sacred sites, so after the floods, I sat by the falls daily. As family members—kids, grandmas, nephews, friends—visited Suzanne's cottage, beginning to manifest her dream of a multigenerational home, a trip to Solstice Falls became an initiation upon each arrival. Before the July rains, this seasonal waterfall had tapered to a June trickle as spring snowmelt became merely a summer memory. But during the floods, the waters surged from the top, cascading onto ancient granite and shale, ripping birch and spruce from their earthen roots while stripping an understory of ferns and loam. A whole new river channel—a minicanyon unto itself—was cut by the surge. The community's trail up Solstice Mountain had disappeared into the rubble of rock pushed from the old river channel.

During a visit just after the floods, my fifteen-year-old nephew, Tyus, and Suzanne's sixteen-year-old son, Cameron, walked to Solstice Falls on their descent from the round, proud peak of Solstice Mountain. Once at the falls, they searched for perfect skipping stones to bring to the lake. Cameron has a vibrant sense of discovery and Tyus feels at home in forests; I thought they might just dig around that shale until nightfall, since the flood had unearthed so much rock—like a found geological treasure chest—unseen for perhaps thousands of years. Suzanne and her eighteen-year-old, Charlie, walked there with Suzanne's mother, Lynn. A volunteer protecting diamondback terrapin habitat back in New Jersey, Lynn shared a sense of excitement for this new, wild habitat that her grandchildren would enjoy for countless years, summarizing the place as "transformative." Charlie thought the light shined on the rock in a way that "made them look soft like wrinkled velvet." Charlie, a violinist and music educator, particularly enjoyed the serene sounds of a smooth cascade flow so recently turbulent with floodwater.

Later, Suzanne, my fifteen-year-old daughter, Atalaya, and I went for a run on a dirt road meandering between Solstice Lake and a preserved forest surrounding Solstice Mountain. My ankle was sore from an old injury. On the jog back, I thought about the waterfall and an ankle plunge cold enough to relieve my pain. On the way to the falls, we saw my mother, Judy—who was also visiting—out walking and invited her along. I knew she would be moved by such a scene so close to Suzanne's cottage. Here, we witnessed the enduring energy of water, tumbling onto streambed rock, echoing off maple and elm trunks. I smiled as I heard my mom's signature, protracted "wow" join these echoes.

As we poked around the base of the waterfall, Atalaya noticed something that summoned in me the same elemental hope I felt when I witnessed Sol pick up her first piece of garbage on the lake. Someone had sculpted an idyllic, circular pool of water out of the rubble recently torn up from the brook. Atalaya and I took off our running shoes and eased our feet into the fifty-five-degree water. It seemed curved perfectly for the human body, as if built for a spa. Never have I felt so invited by the elements.

While Suzanne and my mom pointed out the stunning, wood-land scene to each other downstream, Atalaya and I closed our eyes and named, quietly, everything we heard in this spot. When we really listened, just underneath the sounds of the still-rushing falls, we could hear birdsong, one month after those first birds emerged after the storm. We could hear the breeze whispering through the broad-leafed summer canopy. And what really brought us peace was the steady, slightest trickle of the stream's overflow, continually refilling the pool in which I soaked my ankle. I opened my eyes and saw that Atalaya was still listening with her eyes shut. She and I have done this in so many places that we have a name for it: "taking a picture with our hearts." Atalaya seemed to want to sit there—a teenager with her dad—forever. I know I could have sat forever.

This is the elemental life—hope found from elemental individual, family, and community actions seeking balance with planetary elements, in a place offering simple-yet-stirring human and more-than-human connections. A daughter does not know what to do while grown-ups anxiously struggle with climate chaos, so she starts paddling and cleaning up; a community finds itself without power or accessible roads, so they turn their challenge of no refrigeration into the opportunity to break bread together; two partners building a long-term vision of home seek to invite family into that dream, so they find a shared space of serenity where all can experience "transformation"; devastated multigenerational families need to recharge from resurrecting flooded dreams, so they visit the very headwaters of the problem to find the solution: the element of water, shaped by the element of rock, hugged in air made smoke-free from the storms, fueled and connected and cared for by the whole community.

notes

1. Sean Prentiss, "My Home/It's Called the Darkest Wild," in *What Kind of Ancestor Do You Want to Be?* (Chicago: University of Chicago Press, 2021), 27.
2. Vandana Shiva, "Interview: Vandana Shiva, John Hausdoerffer," in *What Kind of Ancestor Do You Want to Be?* (Chicago: University of Chicago Press, 2021), 190.
3. Charles Darwin, *The Descent of Man, and Selection in Relation to Sex* (N.p.: Digireads. com, 2019), 93.
4. Nel Noddings, *Education and Democracy in the 21st Century* (New York: Teachers College Press, 2013), 118.
5. Aldo Leopold, *A Sand County Almanac* (Oxford University Press, 2020), 104.
6. Leopold, 102.
7. Michael J. O. Pocock, Iain Hamlin, Jennifer Christelow, Holli-Anne Passmore, and Miles Richardson, "The Benefits of Citizen Science and Nature-Noticing Activities for Well-Being, Nature Connectedness, and Pro-Nature Conservation Behaviours," *People and Nature* 5, no. 2 (February 8, 2023), https://besjournals.onlinelibrary.wiley. com/doi/10.1002/pan3.10432.
8. Mirian Barnes, *Care in Everyday Life: An Ethic of Care in Practice* (Bristol, UK: Policy Press, 2012), 6.
9. Lauret Savoy, *Trace: Memory, History, Race, and the American Landscape* (New York: Counterpoint Press, 2016), 2.
10. Prentiss, "My Home," 27.

We Were in a World

Allison Adelle Hedge Coke

We were in a world, in a world, in a world. Sure, we had
our glyphs, but we were providential. Once, some alphabet
believers, glass purveyors, *Ursus arctos* killers, sent all
bailiwick on cursed course far faster gyration backspin,
birling intrinsic angular momentum—boson melts. Spinning,
it careened away iceberg, iceberg, iceberg; glacier braced
time traced yesterday unshakable base—all below flushed
alluvion torrent, Niagara pour, special spate, flux, flow,
until their coastal citadels moldered from cyclone, tsunami,
hurricane gale. Tornadoes tossed turf wherever they pleased.
Eruptions molded Her back into something She deemed
worthy. Not to mention quakes. And the people, the people,
the People, pushed into cataclysm, a few generations from
alphabet book imposed catechism, soon were calamity
tragedy storm splinters, fragmented particles of real past, in a
world gone away from oratory, song, oraliteratures, orations

into gyrations reeling. Soon hot, hot, hot, hot, hot, hot, hot, hot, hot. Hot, dying mangroves, disappearing Waimea Bay, dengue fever, butterfly range shift, meadow gone forest, desert sprung savannah, caribou, black guillemots, bats, frogs, snails—gone. What will sandhill cranes crave? Winged lay early. Reefs bleach. Rain, rain, rain, rain, rain, rain, rain, snow, snow, snow, fires flaming fiercely, fascinated in their own reflecting glare. Marmots rise early. Mosquitoes endure longer, lasting biting spreading West Nile. Polar bears quit bearing. Robins, swallows, enter Inuit life. Thunder finds Iñupiat. Here, it is said, glyphs left rock wall, stone plates, bark, branch, leapt animated into being, shook shoulders, straightened story, lifted world upon their wing bone, soared into Night, to place World back into socket eased sky—stilled us. Some say the soup leftover was worded with decolonized language. Some say the taste lingers even now.

Walking the Elemental World

David Macauley

Elemental Ambulation

I 've experienced some of my most memorable moments when I'm encountering the elemental world on foot, either on regular walks or through long-distance running. I find that walking, in particular, encourages a basic, if often overlooked, form of sensuous engagement with my surroundings, including the capacious earth below, the ambient air above, and the accompanying flows of water along an adjacent path or trail. Through movement across a wide range of terrestrial surfaces, walking stimulates our bodily senses and offers, in turn, a needed counterpoise to both a common cultural tendency to "look down" upon the subtending earth and a corresponding penchant to "fall up" toward transcendental and disembodied heights. In so doing, thoughtful ambulation can contribute to a viable everyday environmental ethic—one whereby we become attentive and sustainable citizens—and transition more easily from our individual bodily sensations to the broader sensuous world itself and eventually to a robust ecological sensibility.

The elemental world—consisting more precisely of the four classical elements of earth, air, fire, and water—is deeply present in many forms of walking, especially when we stretch our environmental imaginations.[1] My walks typically depart from and return to the hearth (fire), the heart of the home. They frequently follow the directional flow and musical rhythms of a river, stream, or seascape (water). They are guided and pleasantly enhanced by the shifting conditions of the sky and atmosphere (air). And they are grounded in and governed by the sensuous surfaces and lay of the land (earth) as well as the confluence of earth and sky, sky and sea, and earth and water in the orientation provided by the encircling or distant horizons.

This elemental fourfold of earth, air, fire, and water is continuously cycled through a local place, gathered, and unified in the motions of the moving body. The American naturalist Henry David Thoreau acknowledged this association when he remarked: "This is a delicious evening, when the whole body is one sense, and imbibes delight through every pore. I go and come with a strange liberty in Nature, a part of herself. As I walk along the stony shore of the pond in my shirt-sleeves, though it is cool as well as cloudy and windy, and I see nothing special to attract me, *all the elements are unusually congenial to me.*"[2] Here, we can see the embodied walker in the depths of an encompassing elemental medium as he detects and enjoys a sense of "sympathy with intelligence" at work in the natural world, where the individual is sensing the surrounding environment but also likely leaving an information trail as he is being watched, felt, and heard himself by other creaturely beings.

We might even distinguish walks through a framework inspired by the four elements themselves and Thoreau's inspired engagement with them. First, there are earth walks, wherein we follow a directional axis across the plane of the unfolding ground, our senses tuned to the sounds, smells, and sights that emerge before us. Thoreau frequently engaged in such sauntering, angling his way into the hills, wandering through the New England woods, and even keeping "appointments" with specific trees he had grown to know and love. The elemental connections to the soil, land, and encompassing place are clearly integral keys to this form of navigation.

Second, we might speak of water walks. In some of his errant outings, Thoreau descended into the moving waters, submerging himself in the palpable thickness of that sensuous element. These "fluvial walks," as he called them, occurred in local creeks, streams, and rivers. In such walks, which the poet William Channing styled "riparial excursions," we literally bathe in the bathos (from the Greek for "depth" and English for "commonplace") of the elemental. We may struggle physically with the force of an element that either resists us—offering its weight in opposition—or, alternatively, that conducts and conveys us when we are walking in the current's given direction. As a Boy Scout, I river walked the streams and waterfalls of Ricketts Glen State Park in northeastern Pennsylvania, mounting fallen tree logs, climbing over outcroppings of rocks, and battling the flows of elemental fluids.

Third, there are sky walks, an idea and practice found among some Native American peoples such as the Tsimshian of the Pacific Northwest, who speak of "one who walks all over the sky."[3] As Thoreau observes: "How few are aware that in winter, when the earth is covered with snow and ice... the sunset is double. The winter is coming when I shall *walk the sky*."[4] Here, we recognize that the air and atmosphere become significant and even visible as cloud, fog, mist, and smoke during a walk to convey an ambient mood and tone, including a sense of time through the position of the sun or filtered light, contributing to the particular rhythms, pace, and tempo of our walks.[5]

Last, we can identify fire walks, even if they should be creatively construed in a more literary rather than strictly literal sense. These "walks" may occasionally involve a negotiation of fiery hot coals, but they increasingly entail the navigation of the lunar landscape (astronomical moon walks), outer space (umbilical space walks), or cyberspace and the electronic ether-world of the internet (virtual walks). They focus on elemental fire—or its domestication through technology—to the extent that they occur outside the sphere of the terrestrial economy of the elements in the thinned ether or ethereal realm of space.

Elemental Appreciation

Admittedly, the elemental world frequently poses resistance and even opposition to our movements on foot; however, such challenges are also potential opportunities to enjoy, understand, and appreciate natural forces and phenomena. The English writer George Meredith, for example, celebrated rain as a "lively companion" in sauntering, one that supplies a secret for intensifying intoxication if one confronts it with love and mirth.[6] Walking in the rain permits the small pleasures of hastening one's step toward a desired destination, relishing the chance to jump puddles, enjoying the suddenly empty trails and streets, or allowing one's clothes and skin to get wet. There is a sensuality associated with walking in the mist, fog, or mizzling precipitation and a corresponding receptivity to the whims of the weather.

While strolling during a rainstorm, we may observe how the sky darkens; how changes occur in the ambient light; how clouds morph, merge, or dissipate; and how water flows, pools, and reflects the sky and passing objects. We witness insects and other animals scuttle for cover or step forth to be cooled or quench their thirst. We might listen to the percussive sounds on the sidewalk, street, or forest floor—the rhythms of dripping, gushing, pitter-pattering, and splashing. We can smell the fragrance of the soil as it is released into the air, a phenomenon known as petrichor. Or we taste water drops as they trickle onto our lips or roll across our tongues. And we may even be rewarded for our patience by the eventual appearance of a rainbow. The human body is, of course, composed greatly of water, and to walk in a rainforest, stroll through an approaching thunderstorm, or hike up a mountain enclosed by clouds is to discover an affinity—and dissolve a sharp distinction—between the inner (corporeal) and outer (ecological) realms.

Walking during a snowfall reminds us that we live within Earth rather than merely residing upon it once we internalize the view that the planet extends beyond the surface to include the encapsulating atmosphere, the medium in which we dwell. Snow and ice, of course, create difficulties for the walker as well as generating beautiful scenes and intriguing sounds. In this regard, the filmmaker Werner Herzog writes of his three-week journey on foot from Munich to Paris, as he walked through an icy world on a lonely winter pilgrimage to visit a dying friend and mentor, offering poetic descriptions of the frozen landscape, unforgiving weather, ecstatic revelations, and physical suffering.[7]

If snow is born of a delicate marriage of water and air, then mud is an untidy admixture of water and earth, a more viscous elemental in-between than ice, for example, which, while likewise composed of water, tends to solidify into a kind of fleeting protoearth. Mud is moist and, as dark pliable matter, usually lacks clear form, making it hard to gain a foothold in and to navigate. Moving through mud on a riverbank, forest floor, swampland, meadow, or bog is either an activity we typically avoid or, alternatively, one we begrudgingly accept along with the reality that it generates shadowy tracks that trail and trace our own forward steps. Thoreau, however, saw the swamp as a sacred site, a *sanctum sanctorum*, and mud as the "marrow of nature." And Gaston Bachelard observed keenly that "walking barefoot in natural, primordial mud awakens our primitive, natural connections with the earth."[8]

While walking almost always occurs within the enveloping medium of air, ambling against the wind can present similar challenges for us, even as it informs us of its invisible strength and reminds us of the larger forces and flows of elemental phenomena. We often lean into the unseen lines of the oncoming wind, pressing our posture against a formidable foe that seems committed to slowing our best efforts, or at least bending or breaking the umbrellas we clutch and carry. Nevertheless, moving air can be a pleasure to feel against our skin and to appreciate when we are warm or exhausted, sauntering on the strand, aware of the drifting scents of flowers and fresh food, or listening to rustling leaves in the forest. As the poet Gerard Manley Hopkins exclaims: "Around, up, above, what wind-walks! what lovely behaviour / Of silk-sack clouds! has wilder, wilful-wavier... I walk, I lift up, I lift up heart, eyes."[9]

Water Walks

As noted earlier, while most walks are oriented by landmarks and the more or less stable earth—trees, mountains, roads, monuments, buildings—water emerges as a guide for many of our outings on foot. The naturalist Edwin Way Teale in fact points out that "the river is the original forest highway."[10] Walking along—or within—streams, brooks, and rivers is especially delightful when the body's rhythms are attuned and even synchronized with the elemental flows of water, contributing to a healthy harmony with the local placescape. "The stream invites us to follow," notes the English writer A. H. Hudson, observing how over time the country annexed the commons and closed footpaths, making it "an offence for a man to go aside from the road to feel God's grass under his feet" and counseling walkers to undress and enter the water if necessary to follow the currents when paths are blocked.[11]

Following a stream, river, or canal is a favorite activity of many naturalists. In a series of meditations entitled *Riverwalking*, Kathleen Dean Moore explores the seductive charms and insights acquired through walks along and into rivers, which she prefers to sitting in a boat, given that such a vessel may separate her too far from the water. "All along the McKenzie River Trail," she writes, "there must be things we do not see, because they have no names. If we knew a word for the dark spaces between pebbles on the river bottom, if we had a name for the nests of dried grass deposited by floods high in riverside trees, if there were a word apiece for the smell of pines in the sunshine and in the shadows, we would walk a different trail."[12] For Moore, wading in and walking through a river allows her to move in close relation to the shimmering reflections of the landscape, while the water bisects her at the waist so that she is both immersed below and observing above the undulating currents.

Natural and anthropogenic coastal lines arise and alight as well through movement by foot along the sea, sand, and shore.

As the philosopher Jacques Derrida once proclaimed, "Everything will flower at the edge."[13] Indeed, we can explore our physical and aesthetic movements in relation to a proliferation of edges and liminal entities at the shoreline, which provides an engaging confluence of earth, water, and sky. For example, we walk to or toward the sea as beachgoers regularly might, or as Gandhi once did more exceptionally in his Great Salt March and pilgrimage to the Arabian Sea. We routinely amble beside the shore on paths, boardwalks, and sidewalks that parallel the seascape. We saunter in and through the cloudy or obscured edges of the ocean itself, which pulses repeatedly against our upright posture. We move out of and emerge from the sea both amphibiously and evolutionarily as beings who remain wedded to earlier biological and watery forms. And on occasion, we even move upon or across the very surface edge of the sea when we stand erect on water skis, surfboards, and the decks of ships, or when we stroll atop floating piers like those created by the artist Christo to enable a kind of walking on water.

The environmentalist Rachel Carson speculates, "To stand at the edge of the sea, to sense the ebb and flow of the tides, to feel the breath of a mist moving over a great salt marsh, to watch the flight of shore birds that have swept up and down the surf lines of the continents for untold thousands of years, to see the running of the old eels and the young shad to the sea, is to have knowledge of things that are as nearly eternal as any earthly life can be."[14] It is here, we might add, that the solidity of the land gives way to the fluidity and liquidity of the sea in a sensuous interstice. The lateral motions and rhythms of our human bodies along the coastline provide a generative contrast to the perpendicular ebb and frictional flow of the incoming and outgoing waters, which themselves are part of a rich panoply and broader scene of alluring edge events and edge entities: roiling waves, fluctuating tides, outcroppings of angular rocks, rims of sand, scuttling creatures, and rugged cliffs. It is here, too, at the edge of the sea that we can creatively engage—through paintings, poetry, and photography, as well as aesthetic perception and embodied participation more generally—aquatic margins, limits, thresholds, and peripheries in their various forms, including the multiple senses of the horizon, along with the anxious status of the beach, which is forever shifting along the coast and beneath our very feet. An illustrious inland example in this regard is Robert Smithson's *Spiral Jetty,* which offers a walkable "pedestrian scale" path of rock and earth in the Great Salt Lake, a watery earthwork even more so than a stable artwork.

In my adolescence, I would walk the beaches of the Gulf Coast of Florida in search of horseshoe crabs—ancient creatures that are not actually crabs but more closely related to spiders—who had been washed ashore by the waves and flipped upside down. I would pick them up, enter the water, and swim out past the line of breaking surf to deposit them in safe territory. In this way, I grew to become more attuned to the beauty and porous intersections of surf, sky, and sand. As Frédéric Gros has noted, "The echoing chants, the ebb and flow of waves recall the alternating movement of walking legs: not to shatter but to make the world's presence palpable and keep time with it."[15]

Elemental Contact and Kinship

By its very nature, walking invites elemental contact and connection. It initiates conversation between our feet and the underlying earth, the encompassing air, the fiery sun, and the circulating waters. It introduces our bodies to and into the natural world, the environing ecological matrix. Ideas and images begin to form and flow; they produce their own kinesthesias of a sort. Our senses even seem to reach toward objects and outlets with a form of projected intentionality. As the poet Rilke puts it in "Spaziergang" (A Walk), "My eyes already touch the sunny hill, / going far ahead of the road I have begun. / So we are grasped by what we cannot grasp."[16]

Touch is the way we sensuously comprehend—literally, hold together in a joint way—the surrounding world. We be-hold place and grow to inhabit it. When we walk within a particular locale, we become an ambling extension of it. We belong to it bodily. The delightful resonance of an echo enchants our ears, and that recoiling voice almost appears to signal poetically that the elements are in accord with us when they reply to our own movements and vibrations. Shakespeare seems to have understood the capacity of the earth to respond and resound when he writes in *Macbeth*'s act 2: "Thou sure and firm-set earth, / Hear not my steps, which way they walk, for fear / Thy very stones prate of my whereabout."[17]

One way to heighten this contact, and hence kinship, with the elements is by walking barefoot, a practice that the novelist John Updike celebrates in a fond recollection of the period he spent at Martha's Vineyard as a youth, indulging in the "lowly element of our Edenic heritage: treading the earth." He catalogs a remembered "medley of pedal sensations," including

> the sandy rough planks of Dutcher Dock; the hot sidewalks of Oak Bluffs, followed by the wall-to-wall carpeting of the liquor store... the hurtful little pebbles of Menemsha Beach... the prickly weeds, virtual cacti... the soft path leading down from this lawn across giving, oozing boards to a bouncy little dock and rowboats that offered another yet friendly texture to the feet; the crystal bit of ocean water; the seethe and suck of a wave tumbling across your toes... the clean wide private sand by Windy Gates... the cold steep clay of Gay Head and the flinty littered surface around those souvenir huts... the startling dew on the grass... the warmth of the day still lingering in the dunes underfoot as we walked back, Indian-file through the dark from a beach party and its diminishing bonfire.[18]

Another way to minimize elemental mediation and thus to enhance bodily contact with the earth, air, light, and water is to reduce the clothing one is wearing or even walk in the nude. In

Germany and Switzerland, for example, some mountain walkers and hikers regularly divest themselves of their apparel and saunter the hillsides, permitting their skin to soak up the sun, feel the wind, and delight in the mist or rain.

Movement by foot across the elemental landscape is often improved when accompanied or guided by seasonal knowledge, too. There are appropriate winter walks, autumn ambles, summer strolls, and spring excursions. Awareness of the seasonal dimensions of bird-watching, medicinal plant gathering, and animal observation—knowing their migrations, hibernations, or breeding seasons—is valuable to the enterprise and practice of learning about the local environment through walking and, of course, coming home to share stories about such journeys.

The changing seasons themselves often encourage us to make connections to the more commodious elemental world. The four (or more) seasons evoke and express elemental oppositions of cold and hot as well as wet and dry. We can loosely correlate winter with earth, which tends to be cold and dry and more exposed and barren of life in this first yearly season. We can reasonably associate spring with water—which becomes wetter and warmer over time—especially as a seasonal catalyst in melting and flooding. We can link summer with fire, which is hot and dry, and particularly the heat and light of the sun. And we might tether autumn with the distinct qualities in the ambient air, which grows cooler and wetter. There are even color correlations that occur as we progress from winter whites and grays to spring greens before moving on to ocean and sky blues of summer, and arriving finally with autumn reds, browns, and gold—an aesthetic point that may interest those of us concerned with fashion and clothing choices.

Elemental Ethics

Generally speaking, the values associated with and created through walking are typically embodied (not only cognitive); they are often pragmatic (not merely theoretical); and they often tend to be democratic and communitarian (as opposed to authoritarian or hierarchical). They are also commonly at odds with those values produced or normalized by an environment organized around auto mobility, energy inefficiencies, and commodity consumption.

The walking figure, too, is routinely a receptive body but also a vulnerable one, and hence his or her world is often in need of being protected. Given the wide variety of ambling types and styles from dog walks to boardwalk promenades, from urban flânerie to wilderness explorations, from solitary strolls to group sauntering, and from political walks to spiritual pilgrimages—walking can help to highlight contrarian, transgressive and eccentric values, too, as well as more collective, ecological, and community concerns. Thoughtful and critical walking practices can deepen an understanding of the elemental environment and, more specifically, strengthen an ethic centered on the land and more-than-human world.

Like gardening, transportation, the dynamics of living space, water and energy use, and cooking or eating, walking is an everyday activity. Thus, ethical questions and issues do not originate theoretically, abstractly, or ideally. They emerge through practice, habit, negotiation, conflict, process, and continual movement, or even intuition in action. They get worked out by being walked out. Walking thus provides a model for refiguring or reflecting upon ethical relationships and responsibilities. It is sustainability through self-transport. Thinking on your feet and often with your feet, so to speak.

Walking offers us a perspective applicable to everyday environmental (and elemental) ethics. By way of analogy, gardening also involves an approach to understanding the natural world. A gardener struggles with weeds, property lines, seeds, soil, fences, compost, and predators. Gardening requires know-how that is learned in the process of actual work. Similarly, the walker struggles with intersections, cars, distances, weather, maintaining attention and focus, sidewalk obstacles, and trail impediments. Wisdom is acquired through work and action rather than mere contemplation; it arrives via engagement as opposed to spectatorship. This demands practical knowledge and practical reasoning— to get from place to place. Theory flows and follows practice; it is built from the bottom upward, as a footing figuratively and literally emerges from the work of the feet.

There are, of course, other ways to create constructive links to the elemental world in addition to mobility on foot or efforts in the garden, including yoga, camping, cooking, photography, and travel. However, walking is a wonderful, easy, and affordable way to commence an everyday elemental practice if one is healthy enough to participate. I invite others to join me on a stroll down the sidewalk, in the woods, along a river, or upon the beach, and to enjoy the simple pleasures, rewards, and kinship of ever-present and enduring earth, air, fire, and water.

notes

1. For my own work on the elements, see David Macauley, *Elemental Philosophy: Earth, Air, Fire, and Water as Environmental Ideas* (Albany: State University of New York Press, 2010).
2. Henry David Thoreau, *Walden and Other Writings*, ed. Brooks Atkinson (New York: Modern Library, 1937).
3. "Tsimshian Legends," Native-Americans.com, https://native-americans.com/category/legends-symbols/tsimshian-legends/.
4. Henry D. Thoreau, *A Writer's Journal*, ed. Laurence Stapleton (New York: Dover Publications, 1960); Thoreau, "A Winter Walk," in *The Writings of Henry David Thoreau* (Boston: Houghton Mifflin, 1906), italics added.
5. See David Macauley, "Head in the Clouds," *Environment, Space, Place* 2, no. 1 (2010): 147–84.
6. George Meredith, "Rain Is a Lively Companion," in *The Vintage Book of Walking*, ed. Duncan Minshull (London: Vintage, 2000), 88–91.
7. Werner Herzog, *Of Walking in Ice* (Minneapolis: University of Minnesota Press, 2015).
8. Gaston Bachelard, *Earth and Reveries of Will*, trans. Kenneth Haltman (Dallas, TX: Dallas Institute, 2002), 101.
9. Gerard Manley Hopkins, "Hurrahing in Harvest," in *Gerard Manley Hopkins: Major Works*, ed. Catherine Phillips (Oxford: Oxford University Press, 2009), 134.
10. Edwin Way Teale, "Walking Down a River," in *The Magic of Walking*, ed. Aaron Sussman and Ruth Goode (New York: Simon and Schuster, 1967), 305.
11. A. H. Hudson, *Afoot in England*, excerpted in *The Vintage Book of Walking*, ed. Duncan Minshull (London: Vintage, 2000), 91.

12. Kathleen Dean Moore, *Riverwalking* (New York: Lyons & Burfurd, 1995), 55.

13. Jacques Derrida, "Parergon," in *The Truth in Painting*, trans. Geoffrey Bennington and Ian McLeod (Chicago: University of Chicago Press, 1987), 81.

14. Rachel Carson, *Under the Sea-Wind* (New York: Penguin, 2007), xix.

15. Frédéric Gros, *A Philosophy of Walking* (London: Verso, 2014), 212.

16. Rainer Maria Rilke, "A Walk," in *Selected Poems of Rainer Maria Rilke*, trans. Robert Bly (New York: HarperPerennial, 1981).

17. William Shakespeare, *Macbeth*, ed. Sylvan Barnet (New York: Penguin, 1998), II.1.

18. John Updike, "Going Barefoot" in *Hugging the Shore* (Hopewell, NJ: Ecco Press, 1994), 61.

Elemental Living:
Elemental Labor and Work

Joerg Rieger

My house is right across a heavily forested area in Nashville, Tennessee, the so-called Warner Parks. Since we can walk there from our front door, my wife and I enjoy the parks several times a week, often hiking but also mountain biking to escape the crowds, take in the clean air, and enter into harmony with the seasons. Up until the COVID-19 pandemic, we were often out there by ourselves. Things changed with the pandemic, as visits to the more than three thousand acres of the Warner Parks went up substantially and increased numbers of people came out to enjoy the woods' almost forty miles of hiking, mountain biking, and horse-back-riding trails, and they kept coming back for many months.

Elementals like earth and air increased in popularity as the pandemic pushed people to realize the profound value of natural areas and breathing fresh air. The ability to move around freely and to breathe deeply in the midst of a global pandemic seemed priceless to many. Fewer people using the parks would have been aware that the land and the forests are also breathing, and that it is that primordial breathing that supports the human ability to breathe. Forests produce oxygen, clean the air in various ways, and absorb twice as much carbon dioxide as they emit.[1] While clean air seemed especially precious during the pandemic, forests functioning as carbon sinks are even more significant at a time when carbon dioxide emissions are the major cause of global warming. In other words, forests and the earth in which they grow are not merely the setting and background of human activity; forests and the earth have their own agency, without which human activity would be endangered and perhaps cease to exist.

Another lesson of the COVID-19 pandemic was a deepened insight into human agency. While human agency does not appear on the list of the elementals of earth, air, fire, and water, its elementary nature finds expression in the term *essential workers*. Just like the agency of earth and forests gained new appreciation during the pandemic, often ignored and undervalued human agency gained new appreciation as well. The agency of essential workers became particularly clear in regard to food production and distribution, health services, and public services. Extending the terminology of the elementals, one might perhaps also speak of the elemental nature of human labor and work. In what follows, I put this observation in conversation with the elemental nature of other-than-human labor and work, as performed by earth, air, fire, and water. Ultimately, it might even be argued that the elementals themselves can benefit from the elemental nature of work sustaining the earth, air, water, and perhaps even fire.

As a first step, consider the essential nature of human and nonhuman work, starting with reproductive labor. Reproductive labor is all work that maintains life and contributes to the flourishing of life. This includes all that is needed for life to emerge, evolve, and sustain itself, often through the menial labor of domestic work and care work but also through wider systems of caring that transcend human efforts. Without reproductive labor, life simply would not exist, and neither would so-called productive labor—the labor that is considered of economic value by producing things for consumption, and which is calculated for gross domestic product (GDP). One example from nonhuman reproductive labor further demonstrates its significance: Without the reproduction of clean air to which forests, oceans, and myriads of bacteria contribute, human life on earth could not exist. However, under the conditions of capitalism, reproductive labor has often been treated as secondary, less valuable than productive labor and for the most part unpaid or underpaid. In capitalist economics, human reproductive labor is thus relegated to females and minority populations, and nonhuman reproductive labor is considered an "externality" whose cost must be kept as low as possible.

Such economic calculations are directly tied to the rampant extraction of environmental resources of the present—often called "raw materials," including fossil fuels—which goes hand in hand with ecological catastrophe. And extraction is linked to pollution, whose consequences are not accounted for in basic economic calculations either. The resulting deterioration and desertification of land, the depletion of valuable resources, and the pollution of air and water (including the vast oceans of the world) are not accidents but linked to the systematic devaluing of reproductive labor in capitalist economies. The dangers of this situation are not limited to hazardous carbon dioxide levels and the lack of clean air but extend to the increasingly limited ability of the land and water to sustain life as we know it.

While each of the four elementals can be understood as performing reproductive labor, consider the contributions of earth and water. As a theologian who is concerned with human relationships to the natural world, I find some wisdom in how different biblical narratives depict the elemental forces of nature. The labor of the elementals of earth and water is an often-overlooked part of the Jewish and Christian creation narratives. In the first chapter of the book of Genesis (1:11–12, 1:20–22, 1:24), earth and water respond to the divine call to bring forth vegetation and living creatures, thereby assuming an essential role in the creation of life. In these ancient stories, no living beings would exist without the agency of earth and water, an insight that is corroborated by modern science as well. Not surprisingly, many contemporary scientific searches for life beyond Earth begin with the search for water on other planets. Another part of Jewish and Christian creation stories—a different literary and theological tradition in the second chapter of the book of Genesis—is Godself engaging the agency of the earth and water by planting a garden next to a stream (2:4–9). In all these examples, ancient religious traditions and modern science agree that various levels of reproductive agency are needed for the emergence and the flourishing of life.

On the basis of these reflections, a deeper understanding of the essential nature of human labor, both reproductive and productive, caring and creative, can be developed. This essential character might elevate it to elemental status, expanding the traditional list of the elementals. No human being (and no mammal in general) would exist without the reproductive gestational labor of mothers. This insight in itself pulls the veil off widely accepted individualist propaganda that promotes the so-called self-made (wo)man. Just like no one can perform the labor of giving birth to themselves, no one is ever self-made and no one can ever pull themselves up by their own bootstraps. New awareness of the essential and therefore elemental nature of labor, starting with gestational labor, has led some feminist thinkers like Elizabeth Freese to argue that prohibiting abortions would amount to sentencing women to forced labor.[2] This perspective might help reshape entrenched positions on the topic of abortion.

The essential and elemental nature of human labor can also be seen in reproductive educational labor in the broadest sense: language would not exist without teaching and learning, as it is not innate in humans, and neither would culture or any of the values considered the bedrock of civilization. While proponents of individualism might not deny these facts, they tend to ignore that these insights extend to all of human life, including the economy, which is built on and sustained by essential and elemental labor. If this is true, the wealthiest persons are not the greatest individuals (as is typically assumed) but the ones most connected and indebted to the reproductive labor of human and nonhuman others. This insight directly contradicts the common assumption that individual hard work is what puts people at the top.

These reflections can be further extended to productive labor. Despite the typical emphasis on the accomplishments of great political or economic leaders, nothing ever gets done without working people. Caesar, for instance, did not build up Rome with his own hands, and not even individual houses would ever get built by contractors and architects without construction workers who perform the essential and elemental bone-breaking labor that often leads to health issues and premature illness and death.

Moreover, images from the Abrahamic religions, which include Judaism, Christianity, and Islam, depict even the Divine as part of the workforce (references can be found in the Tanakh, the Bible, and the Qur'an). Especially in the Tanakh (also known as the Hebrew Bible or Old Testament), God is often portrayed as performing labor, including menial labor and feminized labor. Examples include engagements as craftsman, metalworker, gardener, farmworker, construction worker, shepherd, potter, and garment worker—the latter two tasks mostly performed by women in the societies of the time.

It is striking that all three of the Abrahamic traditions talk about God making Adam from clay (and planting a garden, as mentioned earlier), getting the divine hands dirty, as it were, rather than putting to work migrant workers and artisans, who commonly perform these tasks. The task of the first humans, according to these traditions, is to till and keep the Garden of Eden (Gen. 2:15), creatively combining human and nonhuman labor. Work is not a curse here but part of creation; we might say without equivocation that work is elemental in these traditions. The meaning of another tradition of humanity's "dominion" over the rest of creation (found in a parallel yet distinct story in Genesis 1:26) has long been disputed, as has the notion that a hostile environment would make reproductive and productive human labor more challenging (Gen. 3:15–19). But given the positive references to labor, it does not seem that labor itself is devalued or becomes the problem in any of these traditions.

If labor is indeed essential and elemental, why is reproductive and even much of productive labor valued so little, both human and other than human? As noted, the work of nonhuman nature is generally taken for granted and subject to exploitation and extraction because it is frequently considered an economic externality. In contemporary patriarchal societies shaped by capitalist economies, the work of women and minorities is disproportionally more affected by this devaluation, which is why reproductive labor is typically relegated to them. Even when productive labor is performed by women, in particular agricultural work in many non-Western societies integrated into a global economy, it is because such labor is less valued than the labor of men. Worse yet, some reproductive labor, like the work performed by stay-at-home mothers and some forms of care work in families, is not considered relevant for current economic calculations and therefore is considered to be of no value and is unpaid.

Moreover, even the value of productive labor itself, without which the economy could not function—such as automobile production or construction work—is constantly downgraded. Autoworkers of today often make less than half of what autoworkers made in the recent past (adjusted for inflation), especially when they are not unionized. Other work is downgraded as well, including many of the white-collar professions of the middle class, which is increasingly coming under pressure from developments in artificial intelligence. At the very same time, compensation for corporate leaders keeps rising, so that the gaps in pay are continuing to grow steadily. The CEO-to-worker compensation ratio, which was 399 to 1 in 2021, up from 10 to 1 in 1965, got another boost during the pandemic (even though essential workers were praised) and continues to rise. The Economic Policy Institute identifies the root of this development in the unequal distribution of power between CEOs and working people.[3] What might be done to rectify this?

Karl Marx, drawing on the insights of earlier thinkers, observed the essential and elemental character of work. Following

the insights of the natural sciences of his day, he observed—
within the limits of the binary language of his day—that nature is
the mother and labor the father of wealth. This ties in with our
renewed sense of the essential contributions of elementals like
earth, air, and water to the flourishing of life, derived from the
COVID-19 pandemic. And because nonhuman reproductive labor
is especially devalued in our contemporary economic system,
developing greater respect for the elementals, including the con-
tributions of earth, air, and water, is a great place to start turning
things around. The same is true for developing greater respect for
essential human labor, both productive and reproductive. Again,
the COVID-19 pandemic provided some context for such respect,
even though the practical expressions rarely went beyond applause
for essential workers and other expressions of gratitude that might
have included small gifts, pats on the back, and so on.

In these contexts, developing greater respect for human labor
in all its forms can make a real difference for working people and
their families and communities. In addition, developing greater
respect for nonhuman labor could have tremendous implications
for the appreciation and protection of the environment, positive-
ly affecting life-threatening challenges such as global warming
and growing pollution. Appreciating the life-giving functions of
the Amazon rainforest, for instance, including its role as carbon
sink for the globe and its production of oxygen, healthy soil, and
biodiversity of flora and fauna, would signal the need for limiting
corporate interests that are currently cutting it down to extract its
resources for the production of cattle, soya, pulp (for paper and
biofuel), and mining.[4]

From these observations, developing greater respect for labor,
both human and other than human, can help reconceive the econ-
omy as a whole. Although this is rarely taught in business schools
and even in many economics departments, there are good eco-
nomic reasons for turning things around. Valuing the reproduc-
tive labor of elementals such as earth, air, and water, for instance,

can point economic development toward renewable resources. Electricity produced by wind and water, though not without some shortcomings, helps limit greenhouse-gas emissions and thus global warming. Agricultural production that values and honors the earth has long-term benefits that might make up for the short-term benefits of current models of superexploitation that lead to erosion and loss of valuable resources. And valuing human labor is not simply the decent thing to do (in a situation where a third of children are food insecure even in a highly developed country such as the United States) but makes good economic sense as well, as workers who are appreciated are likely to produce better-quality work and to be healthier and happier. The revitalized labor movement knows this, and its push will make sure that respect for work does not merely remain an idea. Models of worker cooperatives take the value of labor to the next level, completely reshaping our ideas of which forms of labor are essential and which are not.

Finally, developing greater respect for labor, both human and other than human, can also help reconceive culture and even religion. Consider the example of religion. The theologian Paul Tillich defined religion (and the theological study of religion) in terms of what he considered humanity's "ultimate concern." In his magnum opus *Systematic Theology*, he put it this way: "Our ultimate concern is that which determines our being or not-being. Only those statements are theological which deal with their object in so far as it can become a matter of being or not-being for us."[5] While Tillich drew on the categories of existential philosophy to define the ultimate concern, including meaninglessness, anxiety, and despair, reflecting on the elementals and the essential character of work and labor leads to different conclusions about the nature of religion.

A clearer sense of that "which determines our being or not-being" is valuable for a deepened understanding of religion. As noted earlier, the reproductive labor of the elementals of earth, water, and the reproductive work of people—including the gestational work of mothers—are essential for life to exist and to flourish. If this is true, the modern Western assumption that religion is mostly a private and individual matter is faulty. A more comprehensive understanding of religion needs to include what the ancients and many non-Western people knew and know: religion and its practices matters for all of life, and all of life—particularly the various forms of labor—matters for religion and its practices.[6]

These insights help us bring everything together. For the study of religion, this means that matters such as reproductive and productive labor (human and other than human) finally need to be taken seriously, as religion is not primarily a matter of leisure time but of life on the planet as a whole, human and other than human. For the study of politics and economics, this means that essential and elemental matters of life and death need to be studied with much greater seriousness than is currently the case. Reproductive and productive labor—especially the labor of superexploited groups, both human and nonhuman—are not only significant; they are ultimately at the heart of what sustains life on this planet and of how transformation of the status quo might eventually be possible.[7]

All this changes how people engage elementals such as earth, air, and water, and how people engage one another. Elemental living, then, means deepening our sense of what really matters and developing a more profound account of that "which determines our being or not-being." Even a literal "walk in the park," as described at the beginning of this essay, can point us to a fresh understanding of the world and ourselves, which reverses common understandings of what matters more and what matters less. Such deepened understandings, then, have consequences that challenge business as usual under the conditions of neoliberal capitalism, opening new horizons not only for the study but also for the practice of economics and politics, as well as culture and religion.

One of the results of all this, and perhaps the one that matters most at present, when democracy is in danger all around the globe, is a new sense of how far-reaching and meaningful democracy can be. Our sense of democracy, informed by such an elemental perspective, can be expanded from the voting booth (without forgetting the current threats to universal suffrage) to full participation in economics, politics, culture, and religion. The result is democratizing areas of life that are currently devoid of democracy, such as places of work and commercialized earth, water, and air. In this expanded horizon of elemental living, fresh experiences of the flourishing of life should not surprise us.

notes

1. Nancy Harris and David Gibbs, "Forests Absorb Twice as Much Carbon as They Emit Each Year," World Resources Institute, January 21, 2021, https://www.wri.org/insights/forests-absorb-twice-much-carbon-they-emit-each-year.
2. Elizabeth Freese, "The Christian Right's Main Moral Argument against Abortion Rights Completely Ignores This Critical Issue: It's Time to Raise It," *Religion Dispatches*, May 7, 2021, https://religiondispatches.org/the-christian-rights-main-moral-argument-against-abortion-rights-completely-ignores-this-critical-issue-its-time-to-raise-it/.
3. Josh Bivens and Jori Kandra, "CEO Pay Has Skyrocketed 1,460% since 1978," Economic Policy Institute, October 4, 2022, https://www.epi.org/publication/ceo-pay-in-2021/#:~:text=Using%20the%20CEO%20granted%20compensation,the%20composition%20of%20CEO%20compensation.
4. Jonathan Watts, Patrick Greenfield, and Bibi van der Zee, "The Multinational Companies That Industrialized the Amazon Rainforest," *The Guardian*, June 2, 2023, https://www.theguardian.com/global-development/2023/jun/02/the-multinational-companies-that-industrialised-the-amazon-rainforest.
5. Paul Tillich, *Systematic Theology* (Chicago: University of Chicago Press, 1951), 1:14.
6. See, for instance, the work of the Wendland-Cook Program in Religion and Justice at Vanderbilt University, which I direct (www. religionandjustice.org).
7. For further reflections on the relation of religion and productive and reproductive labor in connection with ecological themes, see Joerg Rieger, *Theology in the Capitalocene: Ecology, Identity, Class, and Solidarity* (Minneapolis: Fortress Press 2022), especially chapters 1 and 3.

Elemental Meditation:
Adopting the Pace of Nature
An Interview with Mark Coleman

Gavin Van Horn

*M*ark Coleman is a mindfulness meditation teacher, wilderness guide, and author with over three decades of experience and training in various meditation traditions. Mark leads wilderness meditation retreats from Alaska to Peru, as well as year-long meditation-in-nature teacher trainings in the United States and Europe. His teachings include a focus on the elemental forces of nature and the ways in which those intersect with the understanding and experience of human creatureliness. This interview has been edited for clarity and length.

Gavin Van Horn: The need only seems to be increasing for the kind of guidance and mindfulness work you're doing.

Mark Coleman: I think as the ecological crisis deepens and as people have more stress and are more directly influenced by the impact of ecological disruption, there's more need for teachers and community leaders to be able to provide resources and tools for how to hold eco-anxiety, distress, solastalgia—how to ground, how to feel what's arising through them, and how to respond.[1] So in some ways, that clarifies my work, which is opening people to deeper love of the earth, deeper connection, deeper intimacy, deeper listening, deeper responsiveness. What's required is feeling our love and joy, but also feeling the sadness about what is happening ecologically and finding ways to be present to that.

Gavin Van Horn: One of the things that really stands out to me about your work is that you teach a bodily practice that doesn't merely assuage a mental need; you guide people into a process that they can carry with them and deepen over time. Your *Awake in the Wild* book was where I first encountered your work. When I read the book, I thought, *Finally somebody is taking this practice and not siloing it behind a Zendo or church or meditation walls.* Can you speak to that: why you decided to put the book out and what led you there?

Mark Coleman: I'd been teaching *Awake in the Wild* "Nature in Meditation" work for some years before [the book was published], developing that body of work, my own voice, and my own love and joy and confidence and faith in that work of taking one's meditation outside rather than being in some building, however beautiful the building or temple might be. You know, for so many eons, we've lived and evolved outside. So when I first led a retreat for people in the Southwest in Navajo country, it was so profound and so beautiful, serene and deep. And the wisdom coming from nature, the teachings coming from the natural world, were so palpable, visceral, and immediate. It just felt like a natural place to practice meditation. Why wouldn't you want to meditate and be mindful and contemplative outside? Nature is such a great teacher and mirror filled with so much metaphor and wisdom.

Actually, I wasn't planning to write about it. I got asked to write about it. I didn't see myself as a writer. But it was helpful, as writing is, to formulate ideas and to crystallize what the essence of that work is. And it helped people and provided a doorway to [the experience of], "Oh, there's another way to meditate." Rather than sitting in a closed room with closed eyes and interoceptive attention, nature meditation is being present with senses wide open to really receive and attune to the world around us in beautiful and deep ways.

Gavin Van Horn: I like what you say about senses wide open. My experience with Buddhism has been a more interior practice. It was a welcome shift to have you point in these other directions and remind readers that the Buddha was an outdoor practitioner. He taught in the forest. He taught outside. You're reviving what is essentially a traditional practice. What was your relationship like with nature growing up? Was it important to you?

Mark Coleman: I grew up in northern England on the edge of suburbs bordering farmland. So, a lot of my misspent youth included running around on farms and along the North Umbrian coastline, which is very beautiful, wild, and rugged, and also in woodlands and wherever I could get away from the boredom and tedium of suburban life. My parents were more than happy to kick us out all day in that era when kids just went out and played.

I wasn't conscious of going into nature to have this lovely, beautiful, profound experience. I was just going out as a kid playing. But I have distinct memories of lying down in farmers' fields and watching the wheat and the barley blowing in the breeze, and I was aware that nature provided a lot of peace, ease, stillness, safety, joy, delight, fun, and excitement.

I mean, countless people go to the beach, and they might not ever think of themselves as meditators or spiritual or anything like that, but they might spend a few hours just watching the waves, watching the seagulls, watching the light change and walking along the shoreline, being much more present than they might ever be at home on their phones, watching TV, or whatever. So, there's a way that nature drew me, and I think draws most of us, into this quieter, more contemplative place.

Gavin Van Horn: Were you finding fellow practitioners or mentors resonating with the direction you were heading? Or was there some resistance to you saying, "Hey, we should be spending more time doing *kinhin* [walking meditation] in the garden instead of being on the cushion in the temple or Zendo?"

Mark Coleman: In the beginning, there was not much support or even appreciation. There might have been a little envy because, you know, I was [*laughs*] backpacking in the Sierras, doing rafting retreats down the Green River in Utah, and kayaking in Alaska. And these are all long, silent retreats of seven to ten days, so I think my colleagues didn't really get it. They just thought, "Oh, Mark's into nature and he's a meditation teacher and it's Mark's nature thing," as if it were sort of a quaint idiosyncrasy I was into rather than a deep hunger and longing that's part of our human nature—to want to be outdoors, connect with our environment, connect with the more-than-human world. I don't think my critics realized what a profound support for spiritual practice nature is, for developing understanding, wisdom, awakening, connection, and seeing our place in the web of life.

That was twenty years ago. In the past five years, there's been a significant shift, I think, with the intensifying of the ecological crisis, people's awareness of the preciousness of Earth, and people being so fried with looking at screens and that digital sort of dullness that comes from screen orientation. I think there's been a shift globally—especially I've seen it here in North America and Europe. It's being recognized as a mental health need to go outside, to get sunlight, to get fresh air, to have space and perspective. As that shift happened, accentuated during the pandemic, there's a much deeper appreciation and respect for how important nature meditation practice is as a form that's equally valid [compared] to an indoor retreat in a temple or a Zendo. That's a growing awareness, I'd say, in my Buddhist world and in the meditation communities I'm in.

Gavin Van Horn: You have mentioned the word *present* a few times— as in *being present*—and sometimes that's coupled with being silent.

I wonder if you might say why you think being present is important to our relationships with the natural world.

Mark Coleman: We're in climate catastrophe and disruption because we haven't been present to the natural world in a way that is respectful, intelligent, and understanding. We've related to the earth as a resource to extract and use for our own needs without any awareness or understanding of the impact of those actions. And so, when we are very simply mindful and present—when we go outside—we very easily, intimately, and viscerally feel how interconnected, how interdependent, interrelated, we are with soil and air and light and water and the plant kin and tree kin and bird kin. Being present outdoors wakes up that sensitivity, that intimacy, that knowing. Out of that intimacy and knowing of ourselves as the earth, as nature—not separate from—we see that our actions are crucial.

What we do, of course, impacts everything around us, both personally and collectively. When we are not present, we don't notice what's around us. We don't notice beauty; we don't notice the preciousness of the earth. We don't notice the gifts that we experience every day—the light and the sky and the rain—and so we tend to take for granted what is really not to be taken for granted because it's so amazing that we live on this fertile, fecund, beautiful, abundant earth.

Gavin Van Horn: You mentioned several things that are key to the *Elementals* series. You said *soil* and *light* and *air*. One of the techniques that you use is an "elemental meditation" (see pp. 139-140).[2] Can you say what that is and why that's become something that you've incorporated into your guidance for others?

Mark Coleman: The elements meditation is just one way of many of deconstructing the notion that we are a separate self, that we're a separate identity or entity. If we believe we're a separate identity and the earth is over there and we're not a part of it, then it tends to lead to a much more extractive and exploitative relationship.

Elemental Meditation

In this meditation, you will become attuned to the elements within you, which can help bring a greater sense of interconnection with all life. Begin this meditation by going outdoors. As you take your seat, look around. Notice the trees, grasses, rocks, clouds, and plants. Sense how all life is created by and is an expression of all the elements.

Now close your eyes and sense your body on the earth. Feel the contact with the earth element, the ground under you. Sense how you are part of the earth's surface, seated and aware. The earth under you and the rocks, hills, mountains, trees, and creatures are all made of the same earth element.

Reflect how we all come from the earth: we eat and digest the earth element, we excrete the earth element, and eventually we return to the earth at death. We are never for a moment separate from the earth.

You can sense the earth element directly within you as hardness, density, heaviness, and solidity. Feel now your bones, your knuckles, skull, teeth, jaw, fingers, and ribcage. This hard mass is the earth element. So too are your flesh, organs, skin. Take some moments to feel that. Notice what happens when you sense that the earth element within you is the same element in rocks and mountains. Not separate.

After a few minutes, shift attention to noticing the water element. We are made of more than 60 percent water. The water in the rain, clouds, and lakes is not different from the blood in our veins. Our tears reflect the saltiness of the ocean. We sense water through moisture, sweat, saliva, and fluidity in our joints and in our intestinal tract.

Sense directly the water element within your body. Feel the blood moving as the heart pumps. Notice the wetness of your eyes and mouth. Feel moisture on your skin and fluidity in your joints. The water element outside in clouds and oceans is the same as the water element inside. Not separate.

Now turn awareness to your breath. The air element around you is the same air you inhale. The oxygen in the air is vital to every biological process within you. With each breath, you inhale oxygen released from forests and plankton. Each exhale you breathe out is carbon absorbed by plants and grasses. With each breath, sense how the air element outside and the air element within you is the same element. Not separate.

Sense the warmth in your belly, the heat of your torso and coolness of your skin. That heat in the body is not separate from the fire element of the sun, ninety-three million miles away. We come into this life as a warm-blooded mammal. We retain that heat by digesting and absorbing energy from plants that have harnessed the light of the sun. And when we die, that fire element quickly leaves us. Sense now the warmth of your body and the heat of sunlight. Know how they are connected to the same fire element. Not separate.

Last, be aware of the flow of elements, how they can be felt simultaneously as this flow of inner experience of hardness and density, fluidity and wetness, warmth and coolness, and the inhale and exhale. Know how these are intimately connected to the elements all around you.

After you end this practice, notice the impact of this reflection. See if it changes your sense of connection and belonging as you realize everything around you is made of the same stuff of life.

When we see ourselves as part of the elements or the elements as part of us—the elements that run through every life form, every rock, every stream, every tree, every cloud—if we see that those things that constitute all life also are the fabric of our own cells, then all kinds of things come from that. For one, we feel less alienated, less separate, and we start to feel that we are part of a web that we are affected by and have an impact on.

So, for example, our breath. We breathe; we take in the air element, taking in oxygen released from trees, plants, grasses, plankton. We're intimately in relationship with forests as we inhale. As we exhale that CO_2—so I'm told—if you're sitting in a forest, it is reabsorbed in minutes by photosynthesizing leaves, needles, grasses. We realize we're literally breathing with the plant kingdom, and we are also changing the atmosphere.

You know, we talk about CO_2 levels in the atmosphere as if the atmosphere were something up there, something a couple of miles high. When we exhale, we exhale more CO_2 into the atmosphere. We actually change the atmosphere with each breath. When we realize, "Oh, everything I do has an impact because I am the earth, part of the earth," then that usually leads to "If what I do is connected and affecting other life around me, I need to act with care. I need to act respectfully. I need to realize that when the local mining operation is poisoning the water, or the local tar sands operation is dumping effluents into the groundwater, that's maybe affecting my grandchildren's water supply that I'm currently drinking but may not be drinkable in ten, twenty years." When there's an intimacy like that, it makes it very real—how what's happening around us, what we are doing, and how our society's living is intimately affecting us and all other life forms.

Gavin Van Horn: I wonder if you've ever had somebody say to you, "You know, Mark, it's nice that you are doing this kind of work, but what we really need to be doing is X, Y, and Z." What's your response to people who might view meditation as not active enough?

Mark Coleman: It's a good question. One way to look at that question is to think about what we've been through as a society and as a species in the past three years, right? We've been through a pandemic. Some people may have actually improved their life because they got to work from home and they dropped a commute. But for most people, the levels of stress are increasing—whether it's the pandemic, the economy, or dealing with their eco-anxiety and despair and fear or rage or hopelessness about what's happening ecologically. That's a very real mental health issue, which is only going to get stronger because climate disruption is going to get stronger. Yes, it's true that we need people to be engaged, to be active, to be lobbying our politicians and corporations and changing the systemic structures that are reinforcing and continuing business as usual, which is increasing ecological destruction. But we're also human, and we're feeling creatures. Our bodies and hearts are part of the ecosystem. If this aspect of the ecosystem is not being taken care of, what happens? Well, we could experience things like burnout, despair, rage, anxiety, fear—and when we are gripped in those things, we're just not effective.

Some say, "Well, I need my anger to engage." That can be true. Anger can bring forward an energy to speak out against injustice. But it's not sustainable; if anger is our fuel source, it will burn out. And I've worked with many, many burnt-out activists. We have to take care of our own body and heart—whether that's eating well, sleeping well, exercising well, making sure we have good social connection, making sure the activism we're doing is in community so we're not isolated. And we also need to take care of our mental health. Meditation, mindfulness, taking time to be quiet for a day or a few days is very helpful for our well-being and very helpful for our long-term sustainability.

We talk about the sustainability of ecosystems. Well, we are an ecosystem, and if we just work, work, work, work, work, fight, fight, fight, fight, fight, only focus on negative data, only get angry, we're going to burn out. We're not going to be a sustainable activist. When we can take care of ourselves, particularly whatever difficult emotions are coming up in the face of the climate crisis, we're going to be more effective, less polarized, less jaded, and we're going to have more capacity to deal with the feelings. They're not separate. Dealing with the outer problem is not separate from dealing with the inner. When we take care of the inner, we actually do become a better vehicle for effective action.

Gavin Van Horn: Let's circle back to the elements—earth, air, water, and fire. Can you recall when they became part of your own practice, or when it occurred to you that meditating on and with the elements was an effective way to drop into a deep understanding that we are earthlings?

Mark Coleman: The elements meditation is a sort of classical Buddhist contemplation practice, which is also in the Hindu tradition and in other traditions. I learned it through my own Buddhist training, but I learned it indoors. We're more divorced in a way from the rawness of the elements when we're indoors. When I started meditating outdoors, and particularly when I started teaching this practice, it felt like such a useful frame for helping people see: "Oh, right. When I bring awareness to those elements, I see how profoundly I'm *part of.*"

As I taught it, I also began to feel it more in myself. At different times, different elements have sort of become foregrounded. For example, when I'm camping by a stream for a week up in the mountains, and I'm drinking from the stream or a spring for a week, and I'm sweating and hiking and drinking that water over that week, a good percentage of my body that's water is now the mountain spring. Right? And it's just real. It's not just a nice new idea. The sweat is being replenished by the spring. The water in my body is spring water. That spring water is me; it's not separate.

And then there's deepening my understanding of being of the earth, feeling myself as the earth. Like as I walk, feeling myself as the earth walking. It's not me—as Mark separate, walking on the earth. No. I am the earth walking on itself, feeling itself, sensing itself, knowing itself. So when I'm hanging out with trees and rocks and stones and cliffs, there's a felt sense of "Oh, we're not so different." There's a really profound intimacy that comes from that.

At different times, different elements become more apparent, just as I talked about with the breathing: being aware that when I sit with a tree, I'm literally breathing with a tree. Or that cloud that I'm looking at: I'm going to probably be drinking that cloud that's falling on the watershed here on Mount Tam, which will probably make it into the reservoir and out of my faucet, maybe in a few months. It's real. I'm not looking at the cloud over there as this thing that has nothing to do with me. That's part of all of us.

Gavin Van Horn: Right. It goes back to something you said earlier about the depth of one's care—the stream you can't buy: it is not an abstraction, a blue line on a map. This practice deepens the opportunity for you to care for that stream as a living entity that is not separate from your own identity. The potential for that understanding is increased by focusing on the elemental ways that the stream is a part of you.

Mark Coleman: Sure, sure.

Gavin Van Horn: Would you comment on the deep-time aspect of the elements, or what it means to have a deep-time elemental life?

Mark Coleman: Again, there's something about when we go outside—something that we generally don't really touch into when we're indoors in our homes—where we do, as you say, touch into deep time. Sometimes in the desert I feel that sense of timelessness, or being on an ancient seabed or being beside any mountain range. When I'm in the Southwest and I'm around these canyon walls that are a few hundred million years old, it puts everything into a very, very clear perspective. Or looking up at the sky and the billions of years that we're looking at across time and space. There's something very profound about that perspective. Our ancestors probably didn't know the scientific data on that, but we can't help but feel something of the immensity and vastness and awe and mystery and wonder.

When we step out of the confines of our buildings, homes, and offices, we step into nonlinear time—nonclock time. Nature has its own rhythm—slower. There's a great line from [Ralph Waldo] Emerson: "Adopt the pace of nature. Her secret is patience." I was in the redwoods yesterday, and some of the redwoods are probably a thousand or fifteen hundred years old. Deep time is a different quality of relating to the moment and experience. It's hard to say what that is or put it into words, but we feel it as profound and quieting and deepening something that's a very essential part of human nature.

Gavin Van Horn: I'll give you the last word here about the elements. Is there any story or experience where this kind of elemental perspective was strongly felt by you or somebody who you guided who had the light bulb go on for them?

Mark Coleman: I was recently teaching up in the mountains in northern New Mexico, and this gentleman approached me who had been a Zen monk and living in a Zen monastery for twenty years and had been meditating on the four elements for twenty years in the meditation hall. We were doing the contemplative practice of the elements outside. And he said for the first time *it made sense*—like he'd been doing it for twenty years, and it was just a concept. We know we're made of earth, water, fire, air, and they're in the trees and the rocks and the stones. But to do that outside, where you are feeling your elemental nature, you feel: our bones are not separate from stone, and our blood is not separate from oceans, and our flesh is not separate from soil, and our lungs are not separate from the sky.

That's sobering and mysterious at the same time. And humbling—you know, we're just a bunch of elements that comes and goes [*laughs*]. It comes together for a little while, a few decades, and then we'll scatter, and the heat element will evaporate, and the earth will fall into the soil somewhere, and the water will evaporate. So life goes, and it's beautiful and mysterious in that way. It's a beautiful thing to contemplate.

notes

1. *Solastalgia* is a term coined by the eco-philosopher Glenn Albrecht to identify the grief or longing people experience when confronted by deep ecological changes to places they previously recognized as home. A fuller exploration of the term can be found in Albrecht's *Earth Emotions: New Words for a New World* (New York: Cornell University Press, 2019). "Awake in the Wild" refers to Mark's book, *Awake in the Wild: Mindfulness in Nature as a Path of Self-Discovery,* which is a touchstone for nature meditation techniques and invites readers to develop a loving connection with the earth as a form of environmental activism. Mark regularly leads "Awake in the Wild" mindfulness practices and retreats, as well as in-depth apprenticeship training for those seeking to guide others in contemplative nature practice.
2. This elemental meditation is excerpted from: Mark Coleman, *A Field Guide to Nature Meditation: 52 Mindfulness Practices for Joy, Wisdom & Wonder* (Awake in the Wild Books, forthcoming).

Equinox Ritual with Ravens & Pines

Brenda Hillman

—so we said to the somewhat: Be born—
& the shadow kept arriving in segments,
   cold currents pushed minerals
 up from the sea floor, up through
coral & labels of Diet Coke blame shame
         bottles down there—
    it is so much work to appear!

unreadable zeroes drop lamps
       as mustard fields [*Brassica rapa*]
 gold without hinges, a vital
      echo of caring...On the census,
just write: *it exists!*   Blue Wednesday
      bells strike the air like forks
   on a thrift store plate,
& the shadow moves off to the side...

In the woods, loved ones tramp through
    the high grass; they wait in a circle
      for the fire to begin;
they throw paper dreams & sins upon
         the pyre & kiss, stoking the first
    hesitant flame after touching a match
to the bad news—branches are thrust back
across myths before the flame catches—;
ravens lurch through double-knuckled
    pines & the oaks & the otherwise;
a snake slithers over serpentine
then down to the first
    dark where every cry has size—

(for EK & MS)

To Apprentice an Elemental Life

Carina Lyall

I remove another layer of wool, peeling off my pink, knitted sweater. A gift given to me by my grandma in the 1990s. Twenty-five years on, I'm no longer a teen. It's tight, the sleeves too short, telling a tale of nostalgia and stubbornness. Sweat runs down my spine as my cheeks burn. I put my knuckle in my mouth to soothe the pain a little. A taste of dirt, ash, and blood soaks my tongue. I roll my head around to loosen the stiffness creeping up my neck, shoulders, and jaw. The sense of annoyance sits restless in my arms and I exhale loudly, before I give it another go.

I've been in the firepit—in my teaching space under the soft, green embrace of the beech forest—for quite some time now, with the fire striker (a little spark-making tool forged from different metals and iron) digging deeper and deeper into my hand with every failed attempt to ignite the birch shrub. The skin on one finger has peeled off a little from my hand's slipping and pounding into the earth too many times. Around me, a large group of participants are watching closely. They are here doing a yearlong apprenticeship in nature connection.

The fire pit is on our land, which our landlord refers to as a "very unkept garden." The land is surrounded by agriculture and farmed forest, and beyond that, cultivated field upon cultivated field. Here, however, cocooned by the trees, all of the human-made neatness is absent and we are deep in the wild of it all.

I facilitate and offer the participants a way of working with nature, connection, stillness, answers, peace. "Nature connection courses" is the simplest way to name these practices. In truth, my work for the past two decades has been to explore the story of separation, from earth, body, and one another. Around a fire has been the perfect place to sink into the inquiry of healing that story—personally and collectively. Hundreds of people have come to this humble beech forest with their own longing to unearth something deeply buried in their bodies. They've come to heal a version of the separation story. To discuss and unpack their alienation or disconnection from the natural world, or something else equally important but that remains a wordless mystery. I love these kinds of mysteries. Ungraspable and invisible like air, and still they have the power to move us with great force.

On this day in the apprenticeship, we've come to the module of elements and cycles. Together, we are exploring how these life-giving and life-dependent forces exist around us and within us and yet so much of the time manage to slip through our days unacknowledged. We are unfolding how they weave in and out of our bodies. How they are deeply part of our bodies' construction—breath, bones, blood, spirit. How they guide us and show us that we are in constant interaction with nature around us.

The element of fire is our focus at the moment. The group is waiting for the warmth of the flames, our gathering altar. They are waiting for fire to provide a place to support us in our journeys to connect back into ourselves and one another. A space that invites us to tell stories. To sing. To gaze into the embers and poke in contemplation. Mostly, however, those gathered here are waiting for me to get it started, with high expectations after I pretty cheekily said, "It will only take a few seconds."

But it's not happening.

At some point, lost in being the teacher and in them relying on me to get it done, this struggle to create a fire became a performing situation. Even in the moment, I realized I was locked in a battle to conquer the fire making or show off my experience and skills rather than engage in the act of collaborating with the spirit of fire itself. A relationship that had grown personal and deep over the years. The relationship that I built my work around. But now the facilitator role had whisked in and messed with the humble order of things. To be an apprentice, I was forgetting something crucial and foundational. Something elemental.

My way into nature work developed from a sense of rootlessness and feeling deeply disconnected from people and place. Within my lineage, I have very displaced stories that breathe side by side. The Indigenous ways and the missionary's travels. My father and my mother. Where my heart lived was very different from where my feet kissed the earth.

I have moved so many times it makes no sense to count any-more. Between different countries, cultures, landscapes, soil. The world has been my oyster since I was five years old. The feeling of having no place to plant or claim my roots was the sacrifice of my family's and lineage's nomadic life. And as I grew older, a need to feel a close connection to place started to catch up with me.

Reconnecting to what is elemental turned out to be much more complex and painful than I had imagined. I searched for that connection through people, courses, teachers, and community gatherings, all promising to be the community we were all seeking. I tried to prove my worth, bloodline, or ancestry to those I reached out to, hoping that those people would take me in. Each attempt felt more alienating than the next, because it seemed like so many of us showed up with the same questions, and no one really offered up the answer. Where do you go with these inquiries when you feel orphaned to the world?

On this crooked path, one thing led to another until I stumbled into a course on seasons and elements. The teacher, a soft-spoken woman whose website I can no longer find, talked about the sun and a snake and about being a wild human. At the end of the first—and what would turn out to be the last session—she posted a simple task: to be an apprentice to fire for the next month. That was the only instruction. It was an unusual task—in a world that I've known as hungry for plug-and-play manuals, or scientific proof with reasons we should try a specific practice and a nice list of benefits to seal the deal—to be given such a vague setup for homework. She didn't give us a list of outcomes or illustrations that could guide us. She did smile though.

Even with the confusion and airiness of it all, I was intrigued. I wasn't really sure how to go about it. So the first thing I did was buy a fire striker. Matches felt like cheating. I wanted to find my way by using the most basic tools possible. This desire for basics included a process of trying different time-tested techniques of building fire. The cone and log-cabin shapes became my trusted friends. I spent hours on YouTube watching how survivalists worked their tools. I collected driftwood and fallen branches from the forest behind the house. I gathered herbs to offer the fire in gratitude.

A rich experience that came from my early efforts was understanding that it takes some level of will to make wet wood burn! Yet over time, slowly but surely, your fingers will begin to discern the dry sticks from the newly fallen twigs. Even when they're all a little damp. I spent hours then and since, knees in ash, playing with fire. Bathed in smoke. Trying over and over again. It became my meditation practice. Watching flames die out. Experiencing the satisfaction when I got them to grip a hold of the logs. Sending that fairy-godmother-like woman many thoughts, curses, and love along the way.

This apprenticeship to fire was a very different journey than other education or training routes I've taken. A longer and slower way of learning. We rarely ask inanimate forces, "What are you?" or "What are you asking from me?" because we know that, most likely, we won't get an answer right away. And definitely not in the package we expect.

The course with the woman was supposed to be a yearlong course, but for some reason, it turned out that she did only one session. After that, she seemed to vanish. She never voiced to us as a group something that I've since discovered to be true: you can't be an apprentice to only fire, or the elements, or an ecosystem. They can't be separated in that way. They are interconnected. Entangled in the web of life. She cleverly knew that and left us to discover it on our own. We never received any feedback or updates on what to do next. I suspect that there was no need for anything further, really. We were all left with an open door: be an apprentice.

Not long after, my partner and I took our kids, left the city, and ran for the woods. We had found a run-down, rented house with neighbors you could spot with binoculars. It was south of the city, in a corner of the map we knew little about but felt a strong pull toward. The landlord said we could rent it cheaply if we fixed the place up a bit, and coming from city-priced homes, it felt a little like we'd won the lottery. The house was located at the edge of a forest, and the car had a bumpy climb up a rough driveway, but we were in love and our dreams reached for the clouds. We would soon, to our despair and the local handyman's amusement, learn that fixing a place up required a bit more skill than we had. But the love was pure. As we took root here, driving back to the city to do courses made no sense, so this also became my workplace.

Merging home and work in this way means I spend most of my waking hours here looking out the window at, or sitting immersed in, the landscape. Laying on the ground, eavesdropping on the intense conversations between the wind and leaves or of the magpie who very bravely chases off a buzzard.

Things are slow here. I'm slow. That month with fire changed my body's understanding of time and pace. Being an apprentice became a way of being. The element of fire provided a threshold into the deeper layers of the question I began my journey asking: Where do I belong? This messy place in a forest seemed to welcome an almost-answer to that. A silent whisper: "Maybe here," the trees sang.

I tell myself that I know every inch of this place, the truth being that I've gotten to know only a fraction of it in my nine years of living here. It's not a great or grand place in terms of size or wilderness. It is hard to find that here in Denmark. It's a country with about 2 percent wild nature—or protected land—left. The nature surrounding our home is solely agriculture and farmed forest. Next to that, it's field upon field. I don't call our one-acre patch of yard a garden, because it never really became tamed enough for us to treat it that way. Nettle, roots, and forest animals have had their ways with us, and at some point we put down our shovel and accepted the way things were. There is a wall of thick blackberry bush, shelter to many pheasants, black birds, and things with tails that I get only a glimpse of from time to time. All I know about them is that they like our potatoes. At this point we have decided to take pride in the organically grown vegetables we did succeed with and how they have been feeding the animals of this little forest. All these beings hold the ecosystem here tightly together. Even though the forest has shown no mercy, it does tolerate our stay. Still, I cannot help but ask: What does it do to people and nonhuman kin to live in a place so fully domesticated? How can we be good apprentices to our own wild nature in a place that has been so tamed? Yet I am continuously amazed by the adaptability of the beings living in us all.

I have witnessed elements as they have come into relationship to my body. And I have learned even more as I have witnessed them relate to the land here. Like the undefined wind that takes shape as it entangles itself with the old trees we call the Grandmothers—a few oaks, an ash tree, and some beech trees.

I have learned about earth and how it carries the load and weight of thick root and seeds and is constantly turning, moving, revealing, and doing so in such a subtle way that to the eye it's just still. I have witnessed how we humans like to tie them down into being one thing, one shape, or containing only one metaphorical meaning. Water means flow, air means lightness, earth means groundedness, fire means passion. Then these elements shape-shift and ask me to look again.

The elements surround us—always present, always active. In front of the house, the sea lies as a blanket curving along a very low cliff, coming in soft and clear blue. The water is broken only by a tiny island that belongs to the disappearing 2 percent of wildlands. The island has virgin forest status, which has drawn my curiosity and daydreams to all the wildlife I suspect must be hiding there. A few times we have taken the kayak around the little hill rising up out of the sea. Cormorants rule the sky above and their droppings have killed many of the trees. At dusk, and if you're prone to Hitchcockian inspiration, it's slightly scary scenery.

From the southwest, sweeping the flat hills, the wind rolls over water and most days hits our house with some force. Not much breaks it before it greets us. Living in an old, poorly insulated house with only a simple wood-burning stove as our heat source, we have come to know this wind very well. It's the "put more wool on" wind.

I've spent all these years as an apprentice to this acre of complexity, and I include myself in the elemental cauldron. Observing the elements that seem to never accept the politics of domestication. Observing the breath I've exchanged with the Grandmothers. Watching the plants that have fed us grow, and harvesting when the time was right. Watching the fires lit and the white foam on the waves crashing into the cliff. Watching them each as embodied intelligence and points of views in the world. Amazed by the force and beauty of them all.

Living here has shown me the web of life in the clearest way possible. How we are woven together, dependent on what we can offer each other, human and other than human. That even though we can become infatuated by categorizing and placing things in their own individual boxes, each elemental and being exists in relationship to the rest.

Through these gentle and at times brutal teachers, my roots have grown. They've dug deep into the land without me even noticing, reaching into soil, deep waters, riding the wind. I began to discover through living here that the feeling of orphanhood grabbed me less and less. Even in the darkest of moments, it became hard to further believe that I had ever been abandoned. Because we are always engaging. Breath to breath. Body to body.

Connecting foundational pieces that had nothing to do with being worthy or claimed but everything to do with being home.

I needed that original invitation from the woman who turned me toward an apprenticeship with fire to understand this. I'm still not done forgetting, remembering, and understanding what that means in a life.

Back in the firepit. Drenched in sweat and dealing with a battered ego. I remembered what placed me in the circle of those gathered around me in my forest in the first place. It was not me as the heroine and expert. It was and still is my own interconnectedness more than anything else. It was the questions they brought that were similar to mine. *Will we be claimed? What is our place?*

The little woman's voice echoed in my head. Her simple yet powerful instruction was as clear as a cloud-free, full-moon night. The performer in me withdrew, gave up, let go. Because for every time I stroked the fire striker, it became clearer and clearer to me that I will always need this fire. To live fully, to breathe deeply, I will always need the invitation, just as much as any participant who ever set foot here.

Like them, I will always be an apprentice to this wild elemental life.

Tangled in Bird Play

Priyanka Kumar

We don't stop playing because we grow old. We grow old because we stop playing.
— George Bernard Shaw

W hat could be more elemental than play? One June morning, I was hiking in the Santa Fe National Forest when I came upon an unusual sight. At first, I was delighted to gaze at the pale lemony body of an orange-crowned warbler, *Vermivora celata,* and let out a sharp breath when his rarely seen rust crown came into view. Perched on the tip of a tree limb, the bird then began to spin intently in place. Was this a whimsical gesture? Before my astonished eyes, the warbler took another spin and then paused and spun again. Riveted, I lost myself in the sun-splashed moment. Was the bird engaged in solo play like a young child?

The sighting more than delighted me. It had been a desolate spring, with New Mexicans walloped by historically large wildfires that scorched forests and homes, forcing residents from counties neighboring mine to evacuate. Wildfire smoke drifted into Santa Fe, and I found myself shutting all our windows to keep out the thick, hazy air that stung my nose and eyes. As a naturalist, I relish being outside, tangled in bird calls and migration patterns, or in the luminous blue of mountain bluebirds and whirring hummingbirds charging the air. But now I returned coughing and wheezing from field trips to the Santa Fe National Forest. Soon the forest itself was shut down. During that time, I walked the dirt streets of my neighborhood, skirting bone-dry arroyos. To my dismay, the dry, smoky spring and gusty afternoons kicked up so much dirt and particulate matter that my wildfire cough persisted. Eventually, I was forced to stay shut indoors in a way that I hadn't been even during the peaks of the pandemic.

At long last, the national forest reopened in the midst of surprise June showers, and I eagerly took my husband, Michael, and our two young daughters to a thick-canopied trail with mature pines—one of my birding haunts. The early monsoons had greened the woods, and after weeks of being shut in, I relished hiking among my friends, the old-growth pines, who offered an emerald paradise shimmering with light. As the morning wore on, the sun flared and some raucous flies appeared near areas where muddy rivulets crisscrossed the trail. I was getting away from a cluster of flies when I heard a concerto by the mature aspens.

Eager to spot the singers, I lingered. I grinned when I saw a red-naped sapsucker, *Sphyrapicus nuchalis*, darting among silvery aspen branches, his streaked black-and-white head accented with a bloodred cap and throat. The Cornell Lab of Ornithology describes this bird as an industrious woodpecker with a taste for sugar. How relatable! Spurred by the dramatic sighting, I meandered, letting the music be my guide. The concerto—fluted robin notes layered with nuthatch honks and punctuated by sapsucker drumming— was so seductive that I would abruptly pause to locate the soloists.

It was during such a pause that I noticed a pair of orange-crowned warblers. While the male was spinning in that deliberate way that I found so arresting—I counted eight full spins—a female appeared and flitted about in nearby trees, making *pik-pik* sounds. But the male didn't notice and remained rapt in spin play as light filtering through the canopy aureoled him beautifully. I stood rooted to the spot. In an era of climate change and charred forests, when many of us are reeling from climate anxiety, I drank in succor from this moment of unexpected beauty. My children had walked ahead, with Michael keeping pace, but I lost sight of everything and remained suspended as though the spring air had cocooned the warbler and me in a radiant otherworldly moment.

In July, we drove north to wetter Colorado, where I hoped to get some relief from my stubborn wildfire cough. We planned to stay for ten days in a remote cabin on the edge of a northern Colorado forest. On the way, we stopped at the Denver Art Museum, where our elementary-age children discovered a wide-open play area. In an upstairs lobby, I watched them maneuver a stack of oversized blocks into creative towering shapes that they then began to climb on, giggling as the blocks shifted underneath. They were having fun, suspended in elastic time—one of the gifts of childhood. I didn't want to hurry them. Instead, I pulled up an orange chair and gazed out a glass wall while they went on playing. To my surprise, I found myself daydreaming about the spinning orange-crowned warbler and was drawn further into thoughts of bird play: is play as essential to birds as it is to children?

Sometimes I see my younger daughter break out almost unconsciously into gleeful spins, and, as a child, I too used to spin dreamily in my room. I observe the ways my children play and tussle with their friends, but I also vividly recall the exhilaration of being seven and racing down trails in the foothills of the Himalayas with my friends because we couldn't wait to get to a local festival. As we neared the fairgrounds, we all but flew. We entered the air and soared like chittering sparrows. It was the only time of year when our parents gifted us three sets of clothes—one for each day of the festival—and the sleeves of my new dress fluttered about like wings. At last we alighted near the canvas tents, where a variety of games had been set up for hungry sparrows like us; in one tent, a man rotated a wooden handle attached to a globe, which spun around, and when he stopped, the globe spit out a white ball with a winning number. It was a thrill to win any little prize, but what I wanted most was to remain suspended in this way of being—seamlessly transitioning from one game to another until at last the priest in the most magnificent tent would begin singing to the goddess and we would crowd over and be devoured in clouds of incense, tropical fruits, and chiming bells. The grand finale came when a pyramid of sweets was distributed to all, and the desserts melted in my mouth as deliciously as the play-filled day had whirled by. Our everyday play was quieter, with an old-growth tree trunk serving as the wall of a playhouse, where my brother and his friend fought to the end with the twig soldiers they had made in one "room" while I constructed intricate mud structures, delighting when passing ants wandered in, after which I might decide to take on the twig soldiers, my brother, and his friend in one go.

I wondered whether there was a year when I effectively stopped playing. Was it when, three years later, we moved to Delhi? The urban environment seemingly had clipped my wings, and I could no longer wander as freely as I had done in the wild forested areas of my childhood. Still, I roller-skated like a maniac and, when we moved yet again, took to exploring my neighborhoods on a bike with a banana seat before graduating to street badminton. Did I stop playing, then, upon moving to the West as a teenager? In Toronto, I took up tennis and continued to swim, but something did change. When I look back now, I realize that I was merely trying to *improve* my game; the sense of manic spontaneity—and the joy that springs from it—seemed to fade.

What could be more elemental than play? As children we intuitively know how to play without needing to be taught the skill, and we even conjure toys if we don't have any. But what exactly is play? According to Dr. Stuart Brown, the founder of the National Institute of Play, play isn't frivolous, and it is not just for kids, "but something that is an inherent part of human nature." Brown goes on to define play as a state of mind in which we are "absorbed in an activity that provides enjoyment and a suspension of sense of time."[1]

The word *enjoyment* stems from *enjoir* in Old French, meaning "to make joy" or "take delight in." After mentally unspooling the scene of the warbler spinning, I recalled that not only had the bird been lost in play but also I had grown lost while taking delight in the bird. Could I have been engaged in a form of play while watching the warbler? This felt like an exciting question. Being suspended in time can cause us to enter a dreamlike state that, Carl Jung once wrote, is so flowerlike "in its candor and veracity that it makes us blush for the deceitfulness of our lives."[2]

I suspect that most so-called grown-ups could use more play in our lives. But why do we stop being playful? Because we're too busy surviving? Snared in the busyness of life? Harnessed by dreary hours and digital obligations? Do we risk becoming less than human, elementally human, as we lose our capacity for play? Where is the time, for instance, to experience a playful connection with nature or time for true discovery? Still, if birds who are up against the brute forces of nature can find time to play, perhaps we can also elbow the make-work of our day-to-day lives aside and carve room for making joy. I felt primed to dive deeper into the world of bird play and wondered whether I might also discover insights into how play can have an impact on the lives of humans. Can play, for instance, help us cope with climate anxiety?

Gisela Kaplan, emeritus professor in animal behavior at the University of New England in Australia, divides birds' play behavior into three categories: solo play, such as a single bird skipping or snow romping; object play, in which a bird might carry a stone, drop it, then pick it up again and run with it; and social play, the rarest category, which might involve a bird holding an object in their beak while others chase along.[3]

The study of child development likewise indicates that toddlers often engage in solo play or parallel play when they sit next to other preschoolers but tend to play on their own with objects or toys. As they grow older, children get more adept at social play and enjoy engaging with others their age.

In my years of observing birds, one incident of social play stands out. Michael and I were at the Bosque del Apache Wildlife Refuge in New Mexico watching bald eagles in a cottonwood grove when, slightly farther ahead, we noticed two glistening ravens in the sky playfully heckling a juvenile eagle, *Haliaeetus leucocephalus.* The game went on for some time. We watched in surprise when, instead of becoming defensive, the juvenile eagle in turn began to play with the ravens, executing swooping turns in the air like an amateur pilot.

The common raven, *Corvus corax*, belongs to a highly intelligent class of birds—corvids—and the ones we observed seemed to know what kind of game they were playing while pursuing the juvenile eagle. These ravens and the eagle certainly weren't too busy surviving to play. When young corvids appear to be playful in flight, however, there is an element of functionality to their play: they are often engaged in locomotor play, which also hones their fine motor skills.

A few years back, while observing raptors at the Maxwell Wildlife Refuge in New Mexico, I got close looks at a prairie falcon, *Falco mexicanus*, a magnificent creamy-cacao bird with a telltale malar—mustache—stripe. Later, I was intrigued to learn that prairie falcons are among birds who seem to play. In August 1951, David A. Munro from the Canadian Wildlife Service observed a prairie falcon repeatedly dropping dried cow manure (the size of a robin) in midair and then diving to catch it. Munro understood that play can develop skill in capturing and handling prey objects, but he wondered about "the nature of the stimulus" that caused the falcon to play with an "inanimate and unpalatable object."[4]

I considered the phrase "the nature of the stimulus" and circled back to the question I had posed earlier: What compels birds, or humans for that matter, to play? Is play essential to our well-being? Dr. Brown, the previously mentioned play specialist, and his team spent almost a year interviewing "homicidal males" at the Texas Huntsville Prison, along with a comparable non-incarcerated population in the same socioeconomic class.[5] The research study revealed that the men in prison for homicide had not experienced "rough and tumble play" as children and had no memories of playground buddies or games of chase and escape. It seems that play is a critical component of our overall mental health and life outcome.

A study by Dr. Kaplan suggests that it's play behavior—not tool use—in birds that may be associated with a larger brain and a longer life. "For the past 50 years, international animal cognition research has often related the use of tools such as rocks and sticks to cognitive abilities in animals," she writes. "My study found no significant association between tool use and brain mass. However, very clear differences in relative brain mass emerged when birds showing play behavior were compared to those that didn't play. In particular, birds that played with others (known as social play) had the largest brain mass, relative to body size, and even the longest life spans."[6]

Then it hit me: Isn't creativity a form of play? When I am writing, dreaming up stories, or conjuring onto the page a bird I have seen, am I not engaged in a form of play? Further research led me to the evolutionary development psychologist Peter Gray, who writes: "The characteristics of play all have to do with motivation and mental attitude, not with... the behavior itself. Two people might be throwing a ball... or typing words on a computer, and one might be playing while the other is not."[7]

I'd like to think that when I am typing words, I have a playful attitude. Reflecting on Gray's words heightened my understanding of play to see it not simply as a behavior but also as an approach, an attitude, a way of living. Rather than thinking about play as an age- or species-specific action, it might be useful to consider it as a state of being playful. It was a relief to realize that, even when I left behind my childhood tree-trunk playhouse, I didn't exactly stop playing. Apparently, I still indulge in solo play; it just manifests differently. Albert Einstein famously said that play is the highest form of research. I wonder, however, if we construct too many barriers between work and play or even between schoolwork and play. What if we were to truly approach our work as a form of play? What if we ensured that all children have memories of boisterous play?

On the two-hour drive home from Bosque del Apache, Michael and I found ourselves reminiscing about the juvenile eagle who had refused to be heckled by the ravens and instead began to play with them. The eagle seemed *alive* to the possibility that sometimes the best response to adversity is to chuckle at it. Or simply to play.

In the Santa Fe National Forest, the orange-crowned warbler stopped spinning and then gazed for several moments at the glittering air before at last noticing the female who had been calling. Soon the pair flew through the shadows. I went on hiking along the dappled trail, with swallowtails drifting past, but couldn't get the warbler out of my mind. According to the Cornell Lab of Ornithology, the bird raises his head feathers when excited or agitated; the spinning, however, is not a known courtship display for the bird. W. M. Gilbert and colleagues note that, during courtship, the orange-crowned warbler will "droop wings, spread tail, gape beak, and point head upwards."[8] As I saw it, the warbler had been alive to the sheer radiance of that June morning.

Dr. Kaplan's work suggests that the warbler's spinning and the juvenile eagle's good-sportedness signal intelligence and longevity. Could these findings have implications for how play impacts human brains and life spans? We might not have definitive answers to all these questions yet, but perhaps we might find intriguing clues in our own lives.

To paraphrase the poet Seamus Heaney, sighting birds at play caught my heart off guard and blew it open. They made me recall the girl I once was, who loved nothing better than to race about with friends and wander through emerald trails all day long. I have since tried to give play more room in my life. I can't say that play helped ease my wildfire cough, and it certainly has not eased the global systems causing the fires, but it did help me take this difficulty in stride and allow me to (perhaps playfully) respond to climate disruption with more clarity than anxiety. And while you won't find me spinning on any given day, I am allowing myself to experience birds with unabashed joy.

notes

1. Stuart Brown and Christopher Vaughan, *Play: How It Shapes the Brain, Opens the Imagination, and Invigorates the Soul* (New York: Penguin Random House, 2009), 60.
2. Carl Jung, *Collected Works of C. G. Jung* (Princeton, NJ: Princeton University Press, 1970), 10:134.
3. Gisela Kaplan, "Play Behaviour, not Tool Using, Relates to Brain Mass in a Sample of Birds," *Scientific Reports* 10, no. 20437 (2020): https://doi.org/10.1038/s41598-020-76572-7.
4. David A. Munro, "Prairie Falcon 'Playing,'" *The Auk* 71, no. 3 (1954): 333–35, https://doi.org/10.2307/4081698.
5. Stuart L. Brown, "Consequences of Play Deprivation," *Scholarpedia* (2014): https://doi.org/10.4249/scholarpedia.30449.
6. Kaplan, "Play Behaviour."
7. Peter Gray, *Free to Learn: Why Unleashing the Instinct to Play Will Make Our Children Happier, More Self-Reliant, and Better Students for Life* (New York: Basic Books, 2013), 139.
8. W. M. Gilbert, M. K. Sogge, and C. van Riper, "Orange-Crowned Warbler (*Leiothlypis celata*), Version 1.0," in *Birds of the World*, ed. P. G. Rodewald (Ithaca, NY: Cornell Lab of Ornithology, 2020), https://doi.org/10.2173/bow.orcwar.01.

Invitation

Heather Swan

At the forest cusp
          a sacred pull
from leaf mold,
          xylem, phloem
blood of the sparrow,
          blood of the shrew
all of us of a kind.
          There is an invitation,
beyond the noise
          of industry and ego,
a veil you and I
          can slip through.

Permissions

These credits are listed in the order in which the relevant contributions appear in the book.

"On a Saturday in the Anthropocene" by Elizabeth J. Coleman was originally published in *The Fifth Generation* (Brooklyn, NY: Spuyten Duyvil Press, 2016) and also appeared in *Here: Poems for the Planet,* ed. Elizbeth J. Coleman (Port Townsend, WA: Copper Canyon Press, 2019).

"Imagine We're Ancestors Dreaming" by Sean Hill was commissioned by the Natural Resources Defense Council (NRDC) and published on their website alongside poems by Rachel Eliza Griffiths and Jane Wong as part of a special poetry feature, "Poets Imagine a Different Climate Narrative": https://www.nrdc.org/stories/poets-imagine-different-climate-narrative.

"Supracellular: A Meditation" by Sophie Strand is excerpted and adapted from the forthcoming memoir *The Body is a Doorway: Healing Beyond Hope, Healing Beyond the Human* (New York: Running Press, forthcoming 2025).

Acknowledgments

Our gratitude runs deep for the community of kin who made this series possible. Strachan Donnelley, the founder of the Center for Humans and Nature, was animated and inspired by big questions. He liked to ask them, he enjoyed following the intellectual and actual trails where they might lead, and he knew that was best done in the company of others. Because of this, and because Strachan never tired of discussing the ancient Greek philosopher Heraclitus, who was partial to Fire, we think he would be pleased by the collective journey represented in *Elementals*. One of Strachan's favorite terms was "nature alive," an expression he borrowed from the philosopher Alfred North Whitehead. The words suggest activity, vivaciousness, generous abundance—a world alive with elemental energy: Earth, Air, Water, Fire. We are a part of that energy, are here on this planet because of it, and the offering of words given by our creative, empathic, and insightful contributors is one way that we collectively seek to honor *nature alive.*

A well-crafted, artfully designed book can contribute to the vitality of life. For the mind-bending beauty of the cover design, cheers to Mere Montgomery of LimeRed; she is a delight to work with and LimeRed an incredible partner in bringing to visual life the Center for Humans and Nature's values. For an eye of which an eagle would be envious, a thousand blessings to the deft manuscript editor Katherine Faydash. For the overall style and subtle touches to be experienced in the page layout and design, we profoundly thank Riley Brady. We also wish to thank Ronald Mocerino at the Graphic Arts Studio Inc. for his good-natured spirit and attention to our printing needs, and Chelsea Green Publishing for being excellent collaborators in distribution and promotion.

Thank you to our colleagues at the Center for Humans and Nature, who are elemental forces in their own rights, including our president Brooke Parry Hecht, as well as Lorna Bates, Anja Claus, Katherine Kassouf Cummings, Curt Meine, Abena Motaboli, Kim Lero, Sandi Quinn, and Erin Williams. Finally, this work could not move forward without the visionary care and support of the Center for Humans and Nature board, a group that carries on Strachan's legacy in seeking to understand more deeply our relationships with *nature alive*: Gerald Adelmann, Julia Antonatos, Jake Berlin, Ceara Donnelley, Tagen Donnelley, Kim Elliman, Charles Lane, Thomas Lovejoy, Ed Miller, George Ranney, Bryan Rowley, Lois Vitt Sale, Brooke Williams, and Orrin Williams.

—Gavin Van Horn and Bruce Jennings
series coeditors

John Hausdoerffer offers his deepest gratitude to the authors and poets who contributed to this volume, to the editors and thought-leaders at the Center for Humans and Nature, and to the philosophy program at Western Colorado University. The inspiration for living more elementally comes from his wildly wise daughters (Atalaya and Sol) and his boundlessly thoughtful partner Suzanne.

Contributors • volume v

Allison Adelle Hedge Coke's (she/her) eighteen books include *Look at This Blue* (2022, National Book Award Finalist), *Burn, Streaming, Blood Run,* and *Effigies III*. Following former field-worker retraining in Santa Paula and Ventura in the mid-1980s, she began teaching. She is Distinguished Professor at University of California Riverside (UCR), Director of Writers Week, and Director of the Medical Health & Humanities Designated Emphasis in the UCR School of Medicine.

Elizabeth J. Coleman (she/her) is editor of *Here: Poems for the Planet* (Copper Canyon Press, 2019), an international eco-poetry anthology with a foreword by His Holiness the Dalai Lama, and a Union of Concerned Scientists activist guide. Elizabeth authored two poetry collections from Spuyten Duyvil Press, one a finalist for the University of Wisconsin Press prizes. Elizabeth's translation of *Pythagore, Amoureux* into French was published by Folded Word Press. Her poems appear in numerous journals and anthologies. Elizabeth's most recent collection was a finalist for the 2023 Cider Press Review and 2023 Marsh Hawk Press prizes.

Mark Coleman has practiced Buddhist meditation since 1984. He leads wilderness meditation retreats from Alaska to Peru, integrating mindfulness meditation with nature. Mark is a psychotherapist, life coach, and mindfulness consultant. He is an avid outdoor enthusiast and is passionate about combining the forces of meditation, silence, and nature. Mark's books include *A Field Guide to Nature Meditation: 52 Mindfulness Practices for Joy* and *Awake in the Wild: Mindfulness in Nature as a Path of Self-Discovery*. For more information, visit www.markcoleman.org.

Photo credit: Mel Ota

CMarie Fuhrman (she/her) is the author of *Camped Beneath the Dam: Poems* and co-editor of *Cascadia: Art, Ecology, and Poetry* and *Native Voices: Indigenous Poetry, Craft, and Conversations*. She has published or has forthcoming poetry and nonfiction in multiple journals and anthologies. CMarie is an award-winning columnist for the *Inlander* and director of the Elk River Writers Workshop. She is Associate Director and Director of Poetry at Western Colorado University, where she teaches nature writing. CMarie is the host of *Terra Firma*. She resides in the Salmon River Mountains of Idaho.

Liz Beachy Gómez (she/her) is the author of the book *From the Depths of Creation: A Nature-Based Path to Healing*, published in 2023. She is also the founder of Ewassa, a healing space in the mountains near Denver. Liz has practiced a Brazilian nature-based tradition for many years and is trained in Andean shamanic energy medicine. She draws from both traditions for her healing practice, ceremonies, and courses. Liz holds a bachelor's degree in psychology and anthropology, and a master's degree from the Johns Hopkins University School of Public Health. Please visit ewassa.com to learn more.

David George Haskell (he/him) is a writer and biologist. He is the author of *The Forest Unseen*, *The Songs of Trees*, *Thirteen Ways to Smell a Tree*, and *Sounds Wild and Broken*. He is a William R. Kenan Jr. Professor at the University of the South, a Fellow of the Linnean Society of London, and a Guggenheim Fellow. His writing has been recognized as a finalist for the Pulitzer Prize and PEN/E. O. Wilson Literary Science Writing Award, and winner of National Academies' Best Book Award, Reed Environmental Writing Award, National Outdoor Book Award, John Burroughs Medal, and the Iris Book Award.

John Hausdoerffer (he/him) has been the author or coeditor of numerous books on the intersection of environmental ethics and social justice, including *Catlin's Lament*, *Wildness*, *What Kind of Ancestor Do You Want to Be?*, and *Kinship*. He is Professor of Philosophy at Western Colorado University.

Photo credit:
Geoff Wyatt

Sean Hill (he/him) is the author of two poetry collections, *Dangerous Goods* (Milkweed Editions, 2014), and *Blood Ties & Brown Liquor* (University of Georgia Press, 2008). Hill has received numerous awards, including fellowships from the Cave Canem Foundation, Stanford University, and the National Endowment for the Arts. Hill's poems and essays have appeared in numerous journals, including *Callaloo*, *New England Review*, *Orion*, and *Poetry*, and in over two dozen anthologies. A volume of selected poems has been translated and published in Korean under the title *Crush*. Hill teaches in the MFA program at the University of Montana. Find more information at www.seanhillpoctry.com.

Brenda Hillman's (she/her) eleventh collection from Wesleyan University Press, *In a Few Minutes before Later,* was published in 2023. A recent recipient of the Morton Dauwen Zabel Award from the American Academy of Arts and Letters, Hillman has edited and co-translated over twenty books. A former Chancellor of the Academy of American Poets, Hillman lives in the San Francisco Bay Area, where she is Professor Emerita at Saint Mary's College of California. See http://blueflowerarts.com/artist/brenda-hillman/.

Suzanne Kelly, PhD (she/her), is a writer, farmer, and green burial advocate. She's the author of *Greening Death: Reclaiming Burial Practices and Restoring Our Tie to the Earth* (Rowman & Littlefield, 2015), the recipient of the Green Burial Council's 2015 Leadership Award, and one of the founding members of the Town of Rhinebeck's Natural Burial Ground, the second municipally operated green burial ground in New York State. She lives in New York's Hudson Valley, where she owns and operates a no-till market garden and stewards the Rhinebeck Cemetery.

Photo credit:
Molly Wagoner

Priyanka Kumar (she/her) is the author of *Conversations with Birds*, widely acclaimed as "a landmark book" that "could help people around the world rewild their hearts and souls" (*Psychology Today*). Her essays appear in the *New York Times, Washington Post, Los Angeles Review of Books*, and *Orion*. Her work has been featured on CBS News Radio, *Yale Climate Connections*, and *Oprah Daily*.

For the past 13 years, **Carina Lyall** (she/her) has guided people to come closer to nature. She has years of experience in holding space for transformation. Her focus has been on community building, nature, storytelling, and rewilding in a domesticated world.

David Macauley (he/him) is Professor of Philosophy and Environmental Studies at Penn State University, Brandywine. He has taught at Oberlin, Emerson, and New York University and was a Mellon Fellow at University of Pennsylvania. Macauley is the author of *Elemental Philosophy: Earth, Air, Fire, and Water as Environmental Ideas*; editor of *Minding Nature: The Philosophers of Ecology*; coeditor of *The Seasons: Philosophical and Environmental Perspectives*; and coeditor of *The Wisdom of Trees*. He has published articles on ethics, aesthetics, politics, and Continental thought. Macauley is completing a book entitled *Walking: Philosophical and Environmental Foot Notes* and a collection of parables entitled *Re-storying Wisdom*.

Photo credit:
Margarita Corporan

Matthew Olzmann (he/him) is the author of three collections of poems, most recently *Constellation Route* (Alice James, 2022). He teaches at Dartmouth College.

Yakuta Poonawalla (she/her) was born and raised in India. Her deep love for nature began as a teenager during her first trek in the Himalayas. Since that initiation, she has worked with various nonprofit organizations in India and the United States to inspire and cultivate love and respect for all beings. She currently works with the Golden Gate National Parks Conservancy in San Francisco, California, where she leads community initiatives, creating safe and caring spaces for nature connections. She loves walking in urban and wild landscapes and talking to plants, birds, and butterflies. And she's ever ready to make you a cup of chai.

Joerg Rieger (he/him) is Distinguished Professor of Theology, Cal Turner Chancellor's Chair of Wesleyan Studies, and Director of the Wendland-Cook Program in Religion and Justice at Vanderbilt University. He is author and editor of twenty-six books, including *Theology in the Capitalocene: Ecology, Identity, Class, and Solidarity* (2022), *Jesus vs. Caesar: For People Tired of Serving the Wrong God* (2018), and *No Rising Tide: Theology, Economics, and the Future* (2009). He lectures frequently nationally and internationally, and his works have been translated into Portuguese, Spanish, Italian, Croatian, German, Malayalam, Korean, and Chinese.

Vandana Shiva trained as a physicist at the University of Punjab and completed her PhD at the University of Western Ontario, Canada. She later shifted to interdisciplinary research in science, technology, and environmental policy. Dr. Shiva has contributed in fundamental ways to changing the practice and paradigms of agriculture and food. Her books *The Violence of the Green Revolution* and *Monocultures of the Mind* pose essential challenges to the dominant paradigm of nonsustainable, industrial agriculture. Through her books *Biopiracy, Stolen Harvest,* and *Water Wars,* Dr. Shiva has made visible the social, economic and ecological costs of corporate-led globalization. In November 2010, *Forbes* identified Dr. Shiva as one of the Seven Most Powerful Women on the Globe. Her most recent work, a memoir, *Terra Viva: My Life in a Biodiversity of Movements,* was released in 2022.

Sophie Strand (she/they) is a writer based in the Hudson Valley who focuses on the intersection of spirituality, storytelling, and ecology. She is the author of *The Flowering Wand, The Madonna Secret,* and a forthcoming memoir on disability and ecology: *The Body Is a Doorway.* Subscribe to her newsletter at sophiestrand.substack.com. And follow her work on Instagram: @cosmogyny and at www.sophiestrand.com.

Heather Swan's (she/her) poems have appeared in such journals as *Terrain, Minding Nature, Poet Lore, Phoebe, Raleigh Review, Midwestern Gothic,* and *Cold Mountain.* She is the author of the poetry collection *A Kinship with Ash* (Terrapin Books), a finalist for the ASLE Book Award. Her nonfiction has appeared in *Aeon, Belt, Catapult, Edge Effects, Emergence, ISLE, Minding Nature,* and *The*

Learned Pig. Her book *Where Honeybees Thrive: Stories from the Field* (Penn State Press) won the Sigurd F. Olson Nature Writing Award. She teaches environmental literature and writing at the University of Wisconsin–Madison.

 **Leeanna T. Torres** is a native daughter of the American Southwest, a Nuevomexicana writer with deep Indo-Hispanic roots in New Mexico. She has worked as an environmental professional throughout the West since 2001. Her creative nonfiction essays have appeared in various print and online publications as well as several anthologies.

 **Gavin Van Horn** (he/him) is Executive Editor of Humans & Nature Press Books, the author of *The Way of Coyote*, and the coeditor of *City Creatures*, *Wildness*, and the award-winning five-volume series *Kinship*. He currently resides in the ancestral lands of the Northern Chumash people in San Luis Obispo, California, where you can find him wandering the nearby hills and shores, learning the flowers, trying to go light.